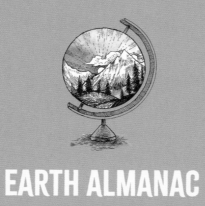

EARTH ALMANAC

EARTH ALMANAC

NATURE'S CALENDAR *for* YEAR-ROUND DISCOVERY

KEN KEFFER

ILLUSTRATIONS BY JEREMY COLLINS

SKIPSTONE

Published by Skipstone, an imprint of Mountaineers Books—an independent, nonprofit publisher
Skipstone and its colophon are registered trademarks of The Mountaineers organization.
Printed in China
24 23 22 21   2 3 4 5 6

Copyeditor: Erin Cusick
Design: Kate Basart/Union Pageworks
Cover and all interior illustrations: Jeremy Collins

Library of Congress Cataloging-in-Publication Data is on file for this title at https://lccn.loc.gov/2019036902.
The ebook record is available at https://lccn.loc.gov/2019036903.

Printed on FSC®-certified materials

ISBN (paperback): 978-1-68051-282-3
ISBN (ebook): 978-1-68051-283-0

Skipstone books may be purchased for corporate, educational, or other promotional sales, and our authors are available for a wide range of events. For information on special discounts or booking an author, contact our customer service at 800.553.4453 or mbooks@mountaineersbooks.org.

Skipstone
1001 SW Klickitat Way
Suite 201
Seattle, Washington 98134
206.223.6303
www.skipstonebooks.org
www.mountaineersbooks.org

LIVE LIFE. MAKE RIPPLES.

TO HEATHER RAY

I look forward to a lifetime of adventures with you,
including regular visits to the oceans and the mountains.
From walks around the block to wilderness treks, I will always remain
by your side, matching your pace, stride for stride. I love you.

CONTENTS

INTRODUCTION

The idea for *Earth Almanac* has been floating around in my mind for years. I was aware of seasonal changes, but it wasn't until I first read *For Everything There Is a Season* by Frank Craighead that I really understood the interconnected ecological relationships of phenology—the seasonal patterns of nature. I soaked it all in, page after page, week by week, summer after summer, as I worked wildlife biology jobs in Grand Teton National Park. My field seasons started as the final snowdrifts melted off in the shadows of the forests. Some years I'd be able to track down a lingering patch of watermelon snow on hikes up Cascade Canyon. Soon, the balsamroot would begin to bloom. The crew all looked forward to grazing in patches of whortleberry by midsummer. I knew the field season was coming to a close when the aspen showed a hint of yellow and the first elk bugles of fall sounded off in the valley.

Embracing the ebb and flow of the seasons is like welcoming a friend back after an extended absence. Each spring the anticipation builds, and we grow eager to see returning migrants or blooms bursting forth. In the dead of summer, we find comfort in knowing the stellar snowflakes are on their way.

Beginning with winter and progressing through the seasons, *Earth Almanac* follows these patterns, exploring the universal flows of nature with more than ninety entries per season. The day-by-day descriptions capture nature's patterns and provide insight into the activities and connections throughout our natural world. While this book can be enjoyed one day at a time, you may find yourself reading it weekly or even seasonally. *Earth Almanac* kicks off with Winter Solstice, the

date on which the seasons pivot, when day length begins to increase and a gradual transition begins once again. Beyond serving as a convenient time to reset your yearly list of species seen, January 1 doesn't mean much to naturalists. It's just another winter day.

Being a naturalist is a mind-set, and *Earth Almanac* unlocks this lifestyle for people who aspire to develop this way of perceiving the natural world. Naturalists are always in tune with their surroundings. They don't look just at birds, mammals, insects, or plants—they take it all in as a whole. *Earth Almanac* embraces this generalist attitude. It provides a solid overview of the outdoors and offers budding naturalists a firm foundation. In some ways, this book is a greatest hits of nature. It highlights many aspects of natural history, including animals, plants, weather, geology, and even astronomy. Fun and enlightening, these nuggets shared over the course of nature's year showcase regional icons and concepts as well as the interrelated connectedness across all of North America between species and abiotic (physical, nonliving) environmental factors. Year after year, this ecological dance coalesces to form the subtle rhythm of Earth.

The tales of nature are far more fascinating than the wildest fiction—and the chance to witness them happens every moment you step outside, and many times when you are stuck inside too. Paying attention to nature allows you to discover daily inspiration in your backyard, when walking your dog to your neighborhood park, while on a hike, or even on a road trip. An awareness of and enthusiasm for nature go hand in hand. Each location hosts its own phenological story. Every spot on the map has a unique natural memoir to share. By choosing to pay attention to these stories, naturalists are "taking the pulse of the planet," according to the National Phenology Network.

An appreciation for nature is only the beginning for some folks. Many wish to give back to the planet by contributing to citizen science efforts. Marked with a citizen science badge, special entries in *Earth Almanac* encourage you to share your sightings to help build a database of observations of nature. Many research projects rely heavily on volunteers to collect integral information about species distribution and abundance. Some citizen science events, such as bird counts and BioBlitzes, include a heavy social component, and other citizen science opportunities can be conducted independently—while you're out on a hike, for example. Another way to give back to the planet is to become a spokesperson for the earth. People are more likely to protect what they are familiar with and what they care about. You don't have to be an expert to share your enthusiasm for nature with others. Be proud of your interests in the outdoors. Wear your binoculars like a badge of honor. Expand your community one person at a time, and you'll build lasting impacts for your neighborhood.

In a world defined by trends, I am happy to report that nature is trending, and it's a trend that is always in fashion. In 1970, Earth Day was founded in part to raise awareness of environmental issues, and it was instrumental in the development of policies like the Clean Air and Water Acts and the Endangered Species Act. These laws are still critically important, but I believe that today's environmental movement brings people back into the realm of nature in a whole new way. And that's a good thing. The health of the earth and the health of people are closely linked. Nature is healthy. A tonic. An anodyne. Time spent in nature is good for the body and for the mind. I encourage you to not only experience the wonders of Earth in the pages of this book, but also to step outside and appreciate our planet firsthand, and as often as possible. There is a solution to nature deficit disorder, as coined by Richard Louv—simply getting outside. If you must give this remedy a name, call it nature immersion therapy.

- MARGARET MURIE GOT IT RIGHT: "Wilderness itself is the basis of all our civilization."
- JOSEPH CORNELL GOT IT RIGHT: Share nature with children.
- EDWARD ABBEY GOT IT RIGHT: "Get out there and hunt and fish and mess around."
- DR. SEUSS GOT IT RIGHT: "Speak for the trees, for the trees have no tongues."
- JANE GOODALL GOT IT RIGHT: "What you do makes a difference"
- SENATOR GAYLORD NELSON, THE FOUNDER OF EARTH DAY, GOT IT RIGHT: Celebrate Earth Day every day.

In the decades since the first Earth Day, we've taken steps as a society to improve how we treat Mother Nature, but the planet still faces numerous demands. You can help by becoming a champion for nature, year after year. From our collective hero Aldo Leopold's early insight into conservation, to the environmental champions of today, we all recognize that it is easy to take nature for granted. It isn't enough to care about nature. It is essential to share nature with other people to help shape their perceptions and actions as well.

WINTER

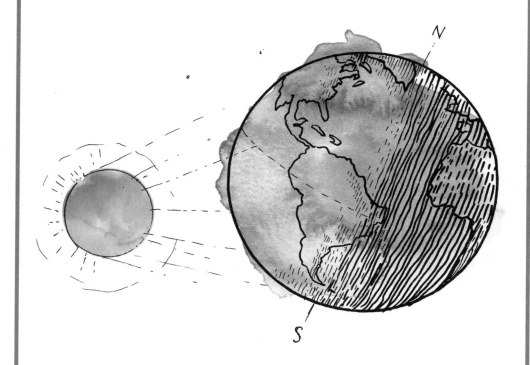

Winter Solstice

WINTER SOLSTICE

The word *solstice* has Latin roots and means "the sun stands still." Cultures have recognized and celebrated the solstice for centuries. In the Northern Hemisphere on December 21, the darkest days of winter are now officially in the past, and starting today, each day brings with it more sunlight than the day before. The tilt of the earth dictates the change in the amount of sunlight from day to day. In the Far North, solstice doesn't look much different than the day before or the day after. Barrow, Alaska, experiences twenty-four-hour darkness from mid-November to mid-January. Fairbanks, Alaska, has three hours and forty-two minutes of sunlight on Winter Solstice.

Why isn't the shortest day of the year also the coldest? The land masses and vast oceans retain the heat they've absorbed during the summer, so there is a delay until the coldest weather of the year hits, usually in mid to late January for much of the United States.

SNOWBIRDS

Many people note seasonal changes by observing the comings and goings of birdlife. Admittedly, it's more difficult to observe them leaving for the season. Did I see the last of the yellow-rumped warblers on Tuesday, or was it three weeks ago? Marking the arrival of a species, however, is easy. Perhaps today the first dark-eyed junco will arrive in your backyard. This species is found year-round in parts of the West, but for folks in the South and East, these birds are the harbingers of winter. Juncos will come to feeders but often prefer to eat seed sprinkled on a platform or directly on the ground.

Sometimes called snowbirds, juncos are fancy sparrows. At one time a number of species were recognized, but now scientists classify the variations as subspecies. Despite plumage variations, all the subspecies have pink bills and white outer tail feathers.

Literal snowbirds, snowy owls and snow buntings also wander south on occasion. A prime location for seeing drifts of wintering snow geese is Bosque del Apache National Wildlife Refuge in New Mexico. Snowy plovers are more at home with beaches and open ground than they are with winter precipitation, so they head to the Pacific and Gulf Coast states for the cold months. Even in the summer, snowy egrets rarely get farther north than southern Canada.

CARIBOU LICHEN BUFFET

Lichens, symbiotic pairings of fungi and algae or cyanobacteria, can grow in crust-like forms on the ground, rocks, and logs, or as branching filaments clinging to trees. One well-known group, collectively referred to as reindeer lichen, is represented by numerous species, especially in circumpolar regions. These lichens are found in a variety of habitats from woodlands to the tundra.

Reindeer Lichen

As the common name implies, reindeer lichen can be an important food source for caribou (another name for reindeer; see September 22). These beasts use their oversized hooves to shovel through snow, digging out lichen to munch on. Icy conditions are difficult to overcome. Caribou also make use of microclimates and find areas in forests with less snow to make foraging easier. Woodland species of caribou eat lichen on trees, and the snowpack helps them reach higher into the branches. Although lichens are a low-nutrition food, caribou have unique symbiotic bacteria in their guts to aid in digestion and nutrient extraction.

HAPPY HOLIDAYS

From Dasher and Dancer to the polar bears and penguins used to peddle soda, the holiday season includes plenty of nature. If you ever encounter reindeer or their free-ranging counterparts, the caribou, up on your rooftop, or anywhere else, listen for the click, click, click. The sound isn't a vocalization; it's caused by tendons clicking as they rub over round sesamoid bones in the animal's feet (similar to your kneecaps, only much smaller). Young caribou don't make the sound when they walk. Some researchers think the sound may help the animals maintain contact with herd mates during dark or stormy conditions.

Caribou are the only members of the deer family in which both males and females sport antlers. Sets of antlers are larger on males, and they shed their headgear earlier in the winter. Females don't lose their antlers until closer to spring. Like many animals, caribou grow thick winter fur coats. Their eyes also change with the seasons, becoming a thousand times more receptive to light in the darkness of winter. Caribou are found throughout the North Pole region, as are polar bears. Penguins, however, are found far to the south.

DECEMBER 25
CHRISTMAS BIRD COUNT

The weather outside may be frightful, but don't just hunker down by the fire. Instead, bundle up and be a scientist for a day. The Christmas Bird Count, administered by the National Audubon Society and Bird Studies Canada, is the longest-running citizen science program in the world. No matter where you live, odds are there is a count circle near you, with more than two thousand in the Western Hemisphere. Even if you don't know an eagle from an egret, you are welcome to join the circle. More eyes looking means more birds spotted. You'll be partnered up with an experienced birder to help you with the identifications in one-day counts during a three-week window in late December and early January. Long-term this data set is helping scientists unravel the mysteries of migration, habitat use, and population trends.

Equally as important, the annual event is a fun tradition for many. At the end of the day, participants in many local circles gather for a tally rally celebration where they swap stories and share some warming food and beverages. Visit www.audubon.org or www.birdscanada.org for more information.

DECEMBER 26
MISTLETOE KISSES

The ancient Greeks and Romans used mistletoe as a cure for a variety of ailments; however, the tradition of kissing under the plant likely dates back to sixteenth-century Celtic Druids. Mistletoe's ability to bloom in winter may have led to it becoming a symbol of vitality and fertility. The European variety was the original mistletoe, but now the name is applied to many different hemiparasitic plants, meaning the plants can photosynthesize but they also rely on a host plant to survive. Witches' broom is another name for mistletoe since it grows in thick clumps along the branches of trees or shrubs.

Mistletoe's growth rarely kills the infested host trees. In fact, forests with high levels of mistletoe host up to three times more cavity-nesting bird species. Wrens, chickadees, and nuthatches all use the tangles for nesting, as do Cooper's hawks and spotted owls. Three species of hairstreak butterflies depend on mistletoe as a host plant. North America is home to more than thirty types of mistletoe. Dwarf mistletoe, a species found mostly on conifers in the West, ejects its seeds from swollen seedpods. Wildlife eat the berries and spread their seeds, but the berries are toxic to humans.

UNDERWATER STARS

Sea stars, a.k.a. starfish, aren't stars. Nor are they fish. Instead they are echinoderms, along with sea urchins (not an unruly child), sea cucumbers (not a vegetable), and sand dollars (not currency either). Over two thousand species of sea stars exist, some with as many as forty arms. They can be found in both tropical and cold coastal regions. Observations of sea stars are critical in helping researchers document the spread of sea star wasting syndrome. Little is known about the disease, but periodic outbreaks occur, infecting sea stars that then succumb to lesions and decay in a matter of days.

Sea stars are also susceptible to predation by birds. Remarkably, some species have the handy adaptation of being able to regrow lost limbs, or the limbs are able to regenerate the central body segment. Sea stars use suction-cup-like tube feet to move about, although they rarely travel far. Sea stars smother their prey, including barnacles, whelks, snails, limpets, and chitons. They also pry open clams and oysters and consume them. Some species of sea stars have specialized stomachs that they can extend outward from their body to envelop and digest food—literally grabbing a bite to eat.

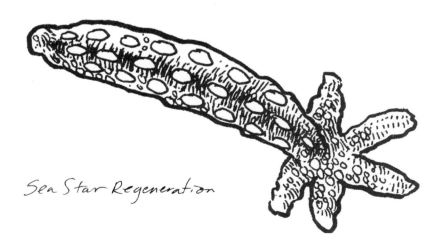

Sea Star Regeneration

SNOW FLEAS

Snow fleas belong to a group of insects called spring-tails, and both descriptive names apply. On warm winter days these arthropods emerge and scatter atop the snow and ice. Lacking wings, the springtail bounces by unclasping its tenaculum, a unique appendage wound up under its abdomen. When they are flitting

DECEMBER 28, 1973
The Endangered Species Act goes into effect. To date, more than 1,400 species have received protection under the law.

about, masses of snow fleas can look like a cloud of pepper shaking on the surface of the snow. Springtails are decomposers that thrive in damp environments. They feed on plant matter, leaf litter, rotting logs, and fungal fluids that they slurp up with modified mouthparts. The species of springtail known as snow flea is most associated with wooded areas and feeds on pollen spores collected on the tops of snowbanks.

MANATEE MERMAIDS

Since sailors far from home often had vivid imaginations, manatees may be the original mermaids. Florida's official marine mammal occasionally goes by the moniker sea cow. It's true these gentle giants are mostly herbivorous, grazing heavily on beds of seagrass, but they are hardly cows. In fact, their closest living relatives are elephants and hyrax. The Florida manatee, a subspecies of West Indian manatee, is the type found in the United States.

Coastal Florida is the core of the year-round range for the manatee. Highly sensitive to the cold, manatees will congregate in relatively warm waters to survive winter conditions. Interestingly, they're able to move freely from salt water to fresh water, allowing them to travel inland quite a distance, especially in winter. The Crystal and Homosassa Rivers on the gulf side of Florida, the Everglades in the south, and Blue Springs on the ocean side are notable areas in their winter range, as are power plant outflows. In summer they can move west to the waters of Texas and along the Atlantic coast to Massachusetts, although sightings anywhere beyond North Carolina are rare. Manatees were never that widespread, but were listed as endangered in 1973. Rebounding populations in the core range led the US Fish and Wildlife Service to downlist them from endangered to threatened in 2017.

HEADGEAR

The terms *antler* and *horn* get tossed around interchangeably, but structurally, they are very different. Covered in nourishing velvet in the summer, antlers are bony appendages that are regrown new each year. Antlers function as displays to attract mates and serve to intimidate rivals. They harden prior to the fall breeding rut, and in winter they're shed, falling off from skull pedicles. Horns, on the other hand, are permanently attached to the animal. They are bony cores covered in sheaths of keratin, and they play a role in social dominance as well as predator avoidance.

One of the defining characteristics between antlers and horns is that only males have antlers (except for caribou), while both sexes may sport horns. For the most part, females generally have smaller and thinner headgear than the males. Here is another way to keep them straight . . . or branched: the forked ones are antlers. Think of the long thin tines of elk antlers. Even the broad flat paddles of bull moose have numerous fingers sprouting from the palms. But take note: Antlers can also grow as single unbranched spikes. Except for pronghorn (see January 15), horns don't split like antlers do, but sometimes they take on different shapes. Bighorn sheep, for example, have horns that curl in a spiral.

GO HOME, BIRD, YOU'RE DRUNK

American Robin

Many berries cling to branches well into winter. Frost-thaw cycles can convert starches in the berries to sugars, which can then ferment into alcohol. When fruit-loving birds like waxwings and robins belly up to the berry bush, it's possible for them to hit the sauce a bit too hard. Reports of drunk birds are rare, but it does happen. The evidence is mostly anecdotal, but young birds appear to have more trouble holding their booze, or more likely, adults tend to avoid the fermented fruits. Alcohol poisoning is a theoretical possibility. Drunk flying could also be a threat, although plenty of sober birds fall victim to flying into objects each year too.

Birds aren't the only ones that will imbibe from time to time. Deer can get tipsy from spending too much time at the orchard bar. And beer mixed with mashed banana is well-known bait for attracting moths to a lit-up bed sheet in backyards in the warmer months (see July 21). Unfermented berries can also cause troubles for the birds. Decorative plantings of nandina, or heavenly bamboo, have been linked to cedar waxwing mortality—one more reason to consider native species for your landscaping needs.

PHENOLOGY

January 1 is the start of a new calendar year, but the date has little impact on the day-to-day life of a hibernating ground squirrel, a nesting great horned owl, or a dormant maple tree. Nature is more in tune with day length, precipitation, and temperature than with the days of the week. Insects emerge in relation to plants leafing out or blooming. Bird eggs hatch when insects are available to feed to growing nestlings. This coming

JANUARY 1, 1970

The National Environmental Policy Act (NEPA) is enacted. The first major law to address the environment, it requires federal agencies to assess environmental impacts before implementing decisions.

and going of the seasons plays out year after year. Phenology is the study of the cyclical and seasonal timing of natural events, the relationship between climate factors and plant and animal life. Weather has impacts in the short term, but long-term conditions play out over the course of multiple years.

Phenological events, like the blooming of flowers or the migration of animals, are sensitive to changes in the landscape. Across the planet, some spring events now occur earlier each year, and fall events happen later and later. These shifts aren't universal, though, and in many cases the direct relationships in this natural balance are falling out of sync. Nature tells a new story each day, culminating in an Earth almanac.

THE HUNTER

Orion, the Hunter, a cool-season constellation, dominates the Northern Hemisphere night sky from January to March. Three stars—Alnitak, Alnilam, and Mintaka—make up Orion's belt. The upper shoulder star is Betelgeuse. At magnitude 0.42 (see February 3), it is the brightest of the stars in the grouping and is categorized by astronomers as a red supergiant. Its red hue is especially evident when viewing it with binoculars or a telescope. On the opposite end of Orion's hourglass body, the blue star Rigel is the constellation's left knee and shines at 0.12 brightness. The fuzzy middle star of the hunter's sword isn't a star at all. It is the Orion Nebula, a birthplace for new stars.

The first known reference to the pattern we call Orion was carved into mammoth ivory. The piece, found in present-day Germany, dates to approximately 35,000 years ago and is attributed to the Aurignacian archaeological tradition.

THE PALMS

Though coconuts aren't native to the United States, the country is home to more than a dozen species of coconut-free palms stretching from the Carolinas to California. Not surprisingly, Florida leads the way in palm diversity. The sabal, or cabbage, palm is the state tree for the Sunshine State. This official designation also applies for the same species in neighboring South Carolina, the Palmetto State. Palm trees aren't trees, however; they are monocots, like bamboo, rushes, grasses, and many flowering plants. In cross section, the trunks of palms are fibrous and somewhat flexible, helpful characteristics for surviving hurricane conditions.

In the southwestern part of the country, the California fan palm is the native species, which is found around the Palm Springs and Twentynine Palms areas. The Rio Grande Valley of South Texas is another palm epicenter. These days, nonnative palm trees line plenty of boulevards and front lawns, but the native palm habitats are unique ecosystems worth seeking out. A lot of critters, including frogs, snakes, bats, and birds, find shelter in the fronds. Small mammals, from mice to coyotes, eat the plant's fallen fruit, which ultimately helps disperse its seeds.

Sabal Palm

MAKING HAY

Pikas resemble small bunny rabbits with stubby ears. They thrive on rocky slopes on scattered mountaintops in the West, from New Mexico to central British Columbia. Perhaps you've heard their sheep-like bleats coming from rock piles along alpine trails. These boulder fields are suitable burrows for the pika, and the nooks also provide pantry space. In territory that's buried under snow for months at a time, pikas eke out a living by storing food. Active herbivores, they spend the short summer collecting a cache of vegetation for the winter. These haystacks, as much as a bushel (about 35 liters) of dried plant matter, provide essential nourishment since pikas do not hibernate.

Though it may seem that warmer temperatures and shorter winters would benefit pikas, instead, the range of the species is contracting. Considered to be climate indicators, pikas are intolerant of temperature extremes. They rely on cool summer temperatures and heavy snowpack to help protect them from the winter cold.

ON THE PROWL

Breeding season for owls heats up around the turn of the new year. Owls can call throughout the year, but they are especially vocal as they establish breeding territories and find mates. You're likely to have at least one species of owl in your neighborhood.

Great horned owls are widespread and live in a variety of habitats. Rarely constructing their own nests, they instead take over structures built by other birds. Great horneds deliver the cliché "hooo, hahoo, hoo, hoo" hoots. Males have larger syrinxes, basically the avian voice box, and deeper calls.

Great Horned Owl Nest

Equally at home in rural and urban settings, eastern screech-owls have high-pitched whinnying trills that sound vaguely horse-like. Western screech-owls utter whistled hoots in an accelerating series of five to nine calls. The most impressive owl cacophony may belong to the barred owl. The "who cooks for you, who cooks for you all" caterwauling can be heard from quite a distance in wooded areas in the East and Pacific Northwest. Bundle up and visit your local nature center for an owl prowl program this season.

FROZEN FROGS

For nearly half the year, wood frogs should be called ice frogs. These forest amphibians have special adaptations that allow them to survive in cold environs. They are the only amphibians that live north of the Arctic Circle, but their range extends south along the Appalachians to northern Georgia and Alabama. While many amphibians hunker down below the frost line, wood frogs winter in leaf litter and can be subject to multiple freeze and thaw cycles. Wood frogs, along with chorus frogs and spring peepers, have unique proteins in their blood and livers that are capable of producing high amounts of glucose and molecules called glycolipids. This adaptation prevents their cells from rupturing when frozen. So when cold temperatures hit, their blood freezes and their bodily functions shut down. As temperatures rise, the frogs thaw, their cells rehydrate, their blood starts flowing again, and they return to an active state.

THE WAX MYRTLE DIET

The names of some things in nature can be downright confusing. The name myrtle, for example, is applied to a variety of unrelated plants. It's also the handle for the eastern type of the yellow-rumped warbler. The myrtle warbler has a white throat, while Audubon's warbler (the western variety of yellow-rumped) has yellow under its bill. In winter, when most warblers are far to the south in Mexico and Central and South America, yellow-rumped warblers are one of the most conspicuous birds along the Eastern Seaboard. Their chirps can be heard constantly at places like Assateague Island National Seashore and Chincoteague National Wildlife Refuge in Virginia and Maryland.

What do they find to eat this time of year? Wax myrtle berries mostly. Despite its name, wax myrtle isn't related to the true myrtle plants of the Mediterranean. To add to the naming confusion, wax myrtle also goes by the name bayberry. The species is common in the southeastern United States, though its range extends from New Jersey to Oklahoma. The salt-tolerant evergreen shrub can form dense thickets behind coastal dunes. Its bluish-purple berries are a food source for wintering birds, and the species is also the host plant for the red-banded hairstreak butterfly.

TALL TAILS

You're probably familiar with the lobster tails of Maine lobsters, the species most folks in North America eat. Also known as American lobsters, this hard-shelled, clawed crustacean lives off the Atlantic coast. The southern Pacific coast hosts the California spiny lobster, a softer shelled, clawless type. Spiny lobsters are often harvested and exported to China for consumption.

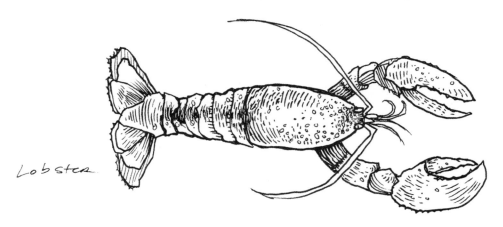

Lobster

All lobsters are omnivores that eat a mix of fresh and scavenged food, including crabs, clams, mussels, worms, sea urchins, and a variety of sea plants. Lobsters are primarily nocturnal, but they have poor eyesight, so they rely on a number of sensory adaptations to explore their surroundings, find their next meal, and avoid becoming prey. Cryptic mottled colorations, coupled with claws or spines, keep them safe from most predators, including cod, flounder, eels, and seals. Harvest limits help protect the lobster fisheries; the smallest and largest individuals are generally returned to the water to grow or breed.

Lobsters molt their exoskeletons about once a year as adults, but larger individuals may go several years between sheddings. Despite internet rumors, lobsters are not immortal. Researchers can determine an individual's age by counting the rings in its eyestalks.

> JANUARY 9

A BLANKET OF SNOW

Although snow is cold to the touch, a blanket of snow provides great insulation. In winter, the air temperature may fall well below freezing, but if you dig down to the ground, you'll find the temperature beneath the snow hovers close to 32 degrees Fahrenheit. Called the subnivean zone, the area under the snow is home to many animals. White weasels (sometimes called ermine) burrow under the snow in search of mice and voles. Many chicken-like birds—including ruffed, spruce, and sooty grouse, as well as willow, rock, and white-tailed ptarmigan—let the snow pile up on top of them during heavy snowstorms. They stay toasty warm under the fresh, white blanket. Winter explorers can use this same insulation to their advantage; snow caves can keep people warm(ish) on a mountainside during winter camping trips or in an emergency situation.

> JANUARY 10

BALMY THROUGHOUT THE YEAR

While the temperature aboveground can fluctuate widely from day to night or from season to season, in the depths of a subterranean cave, it remains fairly consistent. That's not to say there isn't any variation. Cave temperatures are affected by four sources of heat: overlying rock, underlying rock, airflow, and water flow. Temperatures in areas near surface openings swing a bit more than areas deep within the cave system. Throughout most of the four hundred miles of the Mammoth Cave National Park in Kentucky, the average temperature remains near 54 degrees Fahrenheit, but closer to the entry point, it can vary by 10 degrees seasonally. In South Dakota's Wind Cave National Park, the temperature gets increasingly warmer the deeper underground you travel. In the same way that the earth's inner core impacts nearby hot springs, this geothermal gradient influences cave temperatures.

Blanket of Snow

The changes in temperature affect other factors like the evaporation rate within the cave, which alters the formation of stalagmites and stalactites. (To keep these two straight, remember when the "mites" go up, the "tights" go down.) The altered evaporation rate can even change the formations chemically. The mineral aragonite, for example, forms in areas with higher evaporation rates and is found in the popcorn or frostwork formations of Wind Cave.

STICK CAVES

Beavers remain active all year. Well, *active* may be a bit of an exaggeration during winter, but they don't hibernate as commonly believed. Instead, beavers in northern regions tend to spend the cold season hunkered down in their cozy homes, which are generally built of sticks and mud along the banks of ponds or small lakes. These beaver lodges are basically simple man caves without cable TV. The lodge itself provides protection from potential predators, like coyotes, wolves, or wolverines, as well as from the seasonal elements. Inside, the living quarters are cramped, but they keep the occupants safe and dry. Some lodges even have a built-in mudroom or vestibule of sorts near the entry/exit tunnel. But the most important feature of the seasonal lodge is the fully stocked refrigerator.

Beaver Lodge

Beavers spend much of the fall stockpiling food rations for the long winter and harvesting branches and arranging them underwater near the front door. Even if the pond is frozen over, these large rodents thus have access to their food. Lodges are usually, but not always, abandoned during the warmer seasons.

In warmer climates, where animals are less dependent on winter food caches, beaver lodges are less common. Bank dens, however, are used by beavers throughout the range of the species, and females often give birth to kits in the den or lodge. The beaver is the state animal of Oregon, featured on the backside of the flag, and it is also the national animal of Canada.

JANUARY 12

DON'T MIND THE IRRUPTION

The wild cards of bird migration, irruptions aren't the regular seasonal movement patterns we see in most migratory bird species; instead they are sporadic movements that seem to be driven by population dynamics and food availability. Classic irruption species include Bohemian waxwing, pine and evening grosbeaks, the redpolls, and Clark's nutcracker, plus raptors like snowy owl, rough-legged hawk, and northern goshawk. A lack of food is one factor that can drive irruptive migrations. When seed crops are low in the Canadian forests, finches may be seen at backyard feeders across the northern United States.

JANUARY 12, 1995

After being extirpated from the region, eight wolves are reintroduced to Yellowstone National Park for the first time since the 1920s. The wolves were relocated from near Jasper National Park in Alberta, Canada. Six additional Canadian wolves are released later in the month.

An abundant food supply can also lead to an influx of migrants. In some years snowy owls experience unusually high nestling survival rates. This boost in numbers is followed by shifts of young birds moving south. Irruptive migrations are difficult to predict and birds can show up anywhere, which only adds to the anticipation and excitement of seeing an unexpected visitor in your neck of the woods.

JANUARY 13

THE RIPARIAN SOUTHWEST

When conjuring up thoughts of the Southwest, what landscapes do you imagine? Probably not forests of cottonwood trees. These southern woodlands comprise the riparian corridor, or *bosque* (Spanish for "forest"). The floodplain of the Rio Grande is classic bosque with a canopy of cottonwood, mesquite, desert willow, and desert olive. The understory is equally as lush in healthy bosque zones. Seasonal flooding helps shape this landscape, weaving green ribbons of vegetation through the region.

Explore this unique habitat at Bosque del Apache National Wildlife Refuge and Rio Grande Valley State Park, both in New Mexico. In winter, the wetlands of Bosque del Apache are home to thousands of sandhill cranes, snow geese, Ross's geese, and a variety of ducks—species that may seem out of place in the desert. Great flocks fly out at sunrise to feed for the day and return to the wetlands as the sun goes down.

SLIME TIME

Quite large by slug standards, the eight-inch banana slug is one of nature's greatest decomposers. Their core range is in temperate rain forests of coastal Pacific redwood, Sitka spruce, and hemlock from central California to southeast Alaska. Scattered relic populations can be found south to Palomar Mountain in Southern California and inland to the Canadian Rockies.

Banana slugs thrive on forest detritus. Using a radula (imagine a microplane citrus zester), slugs also consume live plant matter and have a fondness for fungi. Slug slime serves many roles for the creatures, including respiration, locomotion, communication, and protection from both predators and dehydration. This mucous is neither liquid nor solid, instead it is termed a liquid crystal because the molecules are arranged in a more orderly fashion than in liquids. During extended dry conditions, the slugs can enter estivation, a hibernation-like state.

Technically hermaphroditic, banana slugs most often end up mating with another slug rather than self-fertilizing. Mating can be a four-hour event and occasionally ends with the female consuming the male's penis. Slugs usually survive this amputation, and in some cases are able to reproduce again.

SPEED GOATS

Lewis and Clark made an astute observation of the pronghorn antelope: "Of all the animals we have seen the antelope seems to possess the most wonderful fleetness." It is true that *Antilocapra americana* is swift, clocking in at over fifty miles per hour. The fastest land animal on the continent today, the species evolved alongside a now extinct cheetah relative. Pronghorn also have impressive vision and can spot danger from quite a distance. They respond by flaring their white rump patches to alert others.

The prong of the pronghorn is found on the bucks, while the female has shorter spike horns, only a couple of inches long at most. Pronghorn are uniquely North American and aren't related to antelopes found in other parts of the world. Two subspecies of pronghorn in the Southwest, the

pronghorn

peninsular and the Sonoran, are endangered. Traveling 150 miles between the Tetons and the Red Desert, pronghorn of western Wyoming undertake one of the longest land mammal migrations in the Lower 48. Herds of hundreds of pronghorn congregate on winter ranges. Their coats are made up of stiff, hollow hairs, which help insulate the animals from winter chill and summer heat.

JANUARY 16

UNDER THE ICE

Aquatic environments feel like an entirely different landscape, even more so when ice shuts out the rest of the world. Whether ice is present or not, water dwellers require the same survival basics as life on land. Phytoplankton and zooplankton often spend winter in the sediment at the bottom of the pond, and many amphibians do the same, entering a hibernation-like state. Fish remain alert throughout the winter, although they are much less active because their metabolic rates slow with the cooler temperatures. Much to the chagrin of hard-water anglers, feeding tapers off in winter too.

For aquatic plants, the darkness of winter is compounded by layers of ice and snow blocking out the sun. The roots remain alive, but many aquatic plants die back each year mainly due to a lack of available oxygen, which can be a limiting factor under the ice (see January 31).

JANUARY 17

THE FATHER OF WILDLIFE MANAGEMENT

In 1933, Aldo Leopold published the first textbook on wildlife management and became the first professor in the newly emerging discipline. Much in the way his scholarly work trained conservation professionals, his now classic tome, *A Sand County Almanac and Sketches Here and There*, has made a lasting impact by

JANUARY 17, 2003
El Yunque National Forest is established in Puerto Rico. It is the only tropical rain forest designated as such.

imparting the "land ethic" to a broader audience. The book was published in 1949, shortly after his untimely death from a heart attack while fighting a grass fire on his neighbor's farm. It documents the coming and going of a year at the Shack, a central Wisconsin farmstead where the Leopold family spent years restoring the woods and prairies. Leopold weaves conservation, policy, and ethics lessons throughout his inspirational essays, each of which connects readers to the natural world and motivates them to treat the land with honor and respect.

RINGS OF LIFE

Trees are unique individuals, and each one has a story to tell. Dendrochronology, the study of tree-ring dating, uses core samples to unravel these tales. Visualize a cross section of a tree trunk. Starting with the outer bark, the layers of a tree include the phloem, cambium, sapwood, and heartwood. Bark protects the tree. Phloem is where nutrients pass through. Cambium is the layer of cells that is actively growing. Water moves from the roots to the leaves through the sapwood via the xylem. Inner sapwood cells die off and form the heartwood, the oldest part of the tree, and they continue to provide support, remaining strong and free of decay as long as the outer layers remain intact.

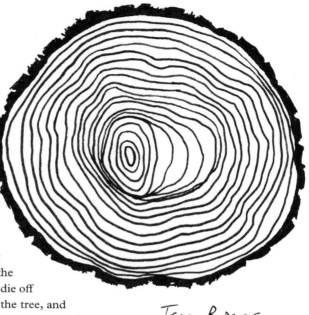

Tree Rings

Dendrochronologists use growth patterns of the wood to piece together the life history of a tree by cross-dating its rings to match specific years. The circular rings are revealed as contrasting light and dark bands. The lighter circles are from larger cells grown in the spring and summer. By late summer, tree growth slows and a darker band of smaller cells is laid down. The inner loops are often wide, reflecting a period of rapid growth in a tree's life. Drought and disease can cause thinner bands. Fire scars leave a signature mark locked in the wood. Asymmetrical growth is indicative of crowding or leaning. In tropical climates growth rings can correlate to the wet and dry seasons, or they can be subtle or even undetectable with the naked eye.

NOMADIC CROSSBILLS

Irruptive migration (see January 12) is an anomaly in the avian world, and crossbills are outliers within this pattern. While most irruptive species show seasonal movements, white-winged and red crossbills tend to roam throughout the year. Especially impressive is that no matter the season, when resources are abundant, crossbills can breed. The birds will nest even in the dead of winter. Dependent on conifer seeds, crossbills have adaptations that help them pry seeds out of the cones, including their odd-shaped bills, the ability to move their upper and lower mandibles from side to side, and elongated tongues they can stick out.

Numerous crossbills can be found throughout the Northern Hemisphere, and their complicated taxonomy continues to be sorted out. The Hispaniolan crossbill, once classified as a subspecies of the white-winged crossbill, is now recognized as a separate species found in Haiti and the Dominican Republic. Similarly, the red crossbill includes numerous types, each with unique calls and bill sizes. The Cassia crossbill of southern Idaho was declared a separate species from the red crossbill in 2017, and ongoing research may one day separate out the other nine types too.

JANUARY 20

UNRAVELING TRACKING MYSTERIES

Animals are always leaving behind clues about the secret lives they lead. Tracks in the mud or snow can tell a story if you know how to read the language. Stride (measured from the heel of one print to the heel of the next) and straddle (measured from the outer width of the right and left tracks) can help roughly determine the size of the critter in question. Stride can help assess speed too. An animal will have longer strides when moving quickly. The shape of a track can also help with identification. Canid (dog relatives) tracks don't always show claw marks. Instead, look for an X pattern extending from the outer foot pad to the inside of the outermost toes. Felid (cat relatives) tracks are generally rounder and have a wider foot pad. The outline of tree squirrel tracks look like small rectangles, but weasel prints can show a similar pattern. A wet substrate, like mud, can make tracks appear larger.

Perfect imprints are rare, but gait patterns can help narrow down species. Sets of rabbit tracks usually show the front feet slightly offset, while the larger hind feet are placed side by side. The trick is that when the rabbit is hopping, its back feet land in front of its front feet. More than just identity, tracks can reveal behavior. Was the animal beelining between two points or wandering? A more sinuous path may indicate active hunting or foraging. The more you observe wildlife in motion, the easier it becomes to visualize the tracks and trails they leave behind. There aren't always definitive answers, but the mystery is part of the appeal of tracking.

JANUARY 21

HOAR FROST

Picture-perfect winter wonderland conditions aren't created by snow at all; it is hoar frost that makes our winter world sparkle. Hoar frost is basically morning dew . . . on mornings when the temperatures may hover in the single digits. The showy, large crystals form as water vapor sublimates onto a surface, say a field of prairie grasses or forest of leafless tree branches. These fragile feathery frosts can be quite photogenic, but they are always ephemeral.

Another type of hoar, depth hoar, is created under a snowpack. Here the water vapor gradient can create layers of large crystals of snow, often called sugar snow. Backcountry skiers dig avalanche pits to help identify these conditions since the depth hoar can create unstable and unsafe situations.

JANUARY 22

WEARING WHITE AFTER LABOR DAY

Fresh snow falls gently to the ground. The flakes pile up higher and higher. Eventually the land changes from stark fall browns to the fresh bright white of winter. Like the landscape, a few animals make this same transition. Weasels, sometimes called ermine in winter, turn white so that they can remain effective predators throughout the year. Weasels in southern states, however, don't turn white for winter, even on the rare occasions when it snows.

Hoar Frost

Snowshoe hares turn white to blend in too, but they do so to avoid being eaten. Researchers found that hares experience a 7 percent higher chance of predation each week that their color mismatches the surrounding habitat. Ptarmigan, chicken-like birds of the north, also molt into white winter feathers. They will bury themselves completely under the snow to stay hidden and warm (see January 9).

JANUARY 23

TALLEST TIMBERS

The tallest trees in the world—redwoods—spire above the landscape in Northern California. Named for their thick red bark, the trees are resistant to fire, insects, and disease. The species can resprout, and this regeneration ability helps the forests thrive. Individual redwoods can live more than two thousand years, although very few of these ancient specimens remain.

Redwoods are as tall as a football field is long. Imagine a thirty-five-story skyscraper, and you'll start to understand the magnitude of a tree that is more than three hundred feet tall. To support this almost unbelievable height, some of the largest redwoods have bases over twenty feet in diameter. They are impressive trees, to say the least. These giants sprout from remarkably tiny seeds, and for such towering trees, the root systems are surprisingly shallow, just ten to fifteen feet deep. Instead, the roots spread out wide, intertwining with neighboring trees for support. The crowns of redwoods are home to entire ecosystems. Redwood branches support plants and wildlife, and the understory is thick with Douglas-fir, western hemlock, rhododendron, ferns, sorrels, mosses, and mushrooms. Huckleberry, blackberry, salmonberry, and thimbleberry are also abundant in the shadows of the redwoods.

The foggy conditions along the West Coast help the trees survive, so redwood forest distribution extends only about 15 miles wide and runs 450 miles long from central California to southern Oregon. These trees occupy just 4 percent of their historic range, with nearly half of existing old-growth redwoods found in Redwood National and State Parks.

JANUARY 24

ITS BILL CAN HOLD MORE THAN ITS BELLY CAN

Winter brings together the two species of pelicans found in the United States, brown and American white. Brown pelicans live on the coasts throughout the year, but these same areas, along with California's Salton Sea, are also the primary winter range for white pelicans. The two species have very different strategies for feeding. Brown pelicans are plunge divers, flying along shallow coastal waters, then diving down from the sky to catch fish they see from above. These dives look violent, even painful, but air sacs under the skin help protect the birds as they crash into the ocean. American white pelicans will forage communally. Multiple birds swim together, circling up fish or pushing them into the shallows to create a brief feeding frenzy.

By February, white pelicans start to migrate north, returning to one of about sixty breeding colonies, including large nesting concentrations at Yellowstone Lake, Wyoming; Chase Lake, North Dakota; and Marsh Lake, Minnesota. Brown pelicans, the state bird of Louisiana, stay and nest in their coastal habitats. The species was in serious peril after experiencing drastic declines due to pesticides but is now a conservation success story as populations continue to rebound.

Brown Pelican

NATIONAL PHENOLOGY NETWORK

 Observing the happenings of nature around you can be a relaxing hobby, but document-ing what you see can also contribute to long-standing research on phenology (see January 1). Nature's Notebook, a program of the National Phenology Network (NPN, www.usanpn.org), lets participants document plants and animals in their own backyards and neighborhoods. Nature's Notebook helps you focus your observations and enhances your appre-ciation for many aspects of certain species. Over 1,300 types of plants and animals are listed to observe. A few focal species are targeted in coordinated campaigns by the organization. Nationwide campaigns include Nectar Connectors, Green Wave, and Pest Patrol. Lilacs are monitored through-out much of the country, while dogwood sightings are helpful in the East. Coordinated by the Uni-versity of Maine Cooperative Extension and Maine Sea Grant with the NPN, Signs of the Seasons is big in New England. The Mayfly Watch program happens throughout the upper Mississippi River and its tributaries. In southern Arizona and New Mexico, flowering agave and saguaro cactus are pollinated by bats, and these blooms receive special attention from researchers.

UBIQUITOUS DANDELION

The dandelion has a lot of haters, but it may be time to wave the white flag and surrender the back-yard to the yellow blooms. Their taxonomy is complex. It is safe to say the most widespread of the weeds originated in Eurasia and is now found in temperate climates worldwide. The common name roughly translates to lion's tooth, a reference to its jagged leaves. If you can find young leaves that haven't been saturated with weed killer, they make a nice, albeit somewhat bitter addition to a salad. Think really assertive arugula. They can also be ground into a pesto. Dandelion wine has only a niche market but can be quite palatable.

Wildlife also make use of dandelions, especially in the early season. It can be one of the first plants available, and nearly one hundred insect species have been recorded utilizing the plant as a food source. At least thirty vertebrates, including numerous birds and chipmunks fresh out of hiber-nation, also feed on the leaves.

MY, WHAT BIG FEET YOU HAVE

Lynx Tracks

Nature provided the inspiration for snowshoes, the footwear that helps us scamper across the snow. Snowshoe hares are built to walk on frozen flakes, as are one of their major predators, Canada lynx. Oversized paws help both species stay on top of the snow instead of sinking in. The dynamics of predator and prey play out over the course of many years for these two interconnected species. When hare populations are high, lynx populations are also strong. The big cats survive by preying on a hare every day or two. The snowshoe hare populations decline cyclically, and lynx populations reflect that pattern. Other northern species like moose don't have the luxury of built-in snowshoes. Instead, these beasts use direct registry to move about in snow, stepping their hind legs into the posthole tunnels that their front feet leave behind. Long, lanky legs help moose high-step through deep drifts. Moose also have another strategy for winter survival: staying put (see February 12).

NOBODY'S HOME

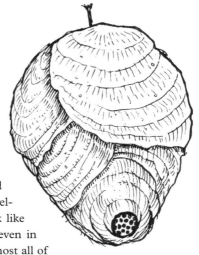

In winter, leafless trees open up views and reveal secrets hidden in the branches since summer. This is the easiest time of year to spot nests in the canopies. Oriole nests can be found dangling down in pendulum-shaped sacks near the tips of branches. Look for a bowl-shaped robin nest placed in the crook of a sturdy branch.

And birds aren't the only nesters. Circular or football-shaped wasp nests can be from bald-faced hornets or one of the yellow jacket species (other types nest underground). They look like nature's piñatas, but there is no need to knock them down, even in winter. These species are generally not very aggressive, and almost all of the individuals die off from starvation when the cold sets in. Only mated

Bald-Faced Hornet Nest

queens survive, usually under the shelter of tree bark. In the spring they will initiate the construction of a new nest. Other species of paper wasps build nests that are shaped a bit like cocktail umbrellas. Each of the open cells in these nests holds a single egg. Wasp nests are constructed from regurgitated bits of plants.

DRAB FINCH

Lemon-yellow American goldfinches are popular backyard birds. Even if you have goldfinches around the neighborhood, you may not recognize them in winter. Unique among the finches, they undergo a second molt of feathers each fall as they slowly transform to a more subdued brownish with hints of yellow, starkly contrasting with their breeding plumages. Females are subtler in summer, but they, too, undergo seasonal molts. For extra insulation, the winter molt includes nearly twice as many feathers as the summer. By March, gold feathers begin to sprinkle back into the plumage, but for now, you're looking for drab finches. American goldfinches prefer eating at thistle feeders but will also consume black oil sunflower seeds.

JANUARY 29, 1927

Edward Abbey is born in Indiana, Pennsylvania. Much of Abbey's writing focuses on the Southwest, including *Desert Solitaire*, about his time as a park ranger in Arches National Park, and the novel *The Monkey Wrench Gang*, which follows a hodgepodge crew of environmental protectors.

The species is the state bird of Washington, Iowa, and New Jersey. In the nation's middle latitudes, the American goldfinch is found year-round. They exit the upper Great Plains and southern Canada in winter, and streaked pine siskins and redpolls often travel with goldfinches in the north. The Southwest has two additional species of goldfinch: the lesser goldfinch and Lawrence's goldfinch. Lesser goldfinches reach the Black Hills of South Dakota in summer but spend winter along the Pacific coast and throughout the southwestern states. This species is widespread in Mexico and south to Peru. Lawrence's goldfinches overwinter on both sides of the Rio Grande from California to West Texas. Breeding takes place in the woodlands of California and Baja California, but sites vary from year to year.

POO-POO PLATTER

Unless you're a fly or a dung beetle, the act of eating feces can seem off-putting. But in the animal world, this practice can be beneficial in several situations. Coprophagy, feeding on dung, can include eating one's own excrement. The lagomorphs, including rabbits, hares, and pikas, are perhaps the most recognized for this behavior. They produce two different types of poo: Hard pellets are their traditional scat, and they also release a soft pellet called a cecotrope. These turds are produced in the cecum, a pouch of the large intestine, and after the animals eliminate them, they reingest the cecotropes for a second digestive pass to extract additional nutrients.

Ruminants, including deer, elk, and bighorn sheep, accomplish a similar double digestion by regurgitating and chewing a cud rather than eating feces. Other examples of coprophagy include the young of some species, such as rabbits, ingesting fecal matter from their mother as a way to boost the gut bacteria required for digestion.

JANUARY 31

WINTER SUFFOCATION

The amount of dissolved oxygen in water is in a constant state of flux. Moving water absorbs more oxygen from the atmosphere than stagnant water does. While cold water generally holds more dissolved oxygen than warm water, when a body of water is covered in ice, oxygen exchange is greatly reduced. Aquatic critters are usually less active in winter and require less oxygen. But long periods of frozen conditions can lead to anoxic conditions. Some species like catfish and pike can handle these extreme low-oxygen conditions better than others. Trout are especially sensitive to low oxygen amounts. High mortality rates can occur if the low-oxygen levels are prolonged. Evidence is found in the spring with high fish kills in some places.

Like pike and catfish, many reptiles, including snapping turtles and painted turtles, are able to cope with low-oxygen levels especially well. They use lungs to breathe in the summer, but when submerged for winter, they get oxygen directly through their skin, mouth, and cloaca. The turtles also break down glycogen in their bodies, which can cause lactic acid to build up, but they balance it by pulling calcium from their shells.

FEBRUARY 1

FRINGED TOES

Ruffed grouse are built for winter. In fall they grow fleshy projections along the lengths of their toes. This fringe increases the surface area of the feet, and some evidence suggests that it may function as a snowshoe for the birds. Another theory gaining more traction is that these pectinations are like crampons that allow the birds to perch on icy branches more easily. Ruffed grouse spend considerably more time in trees in winter than in other seasons because their diet shifts to predominately deciduous tree buds, twigs, and catkins. These prominent perches are exposed to predators, but ruffed grouse are able to dine and dash. In about twenty minutes they can nibble off a day's worth of food, storing the morsels in their crop. When they settle in to consume this takeout, a strong gizzard allows grouse to break down the woody fibers.

Ruffed grouse will burrow under a snowy blanket, but if there isn't enough powder, the birds seek out thermal microclimates in dense stands of conifers. The birds also grow additional feathers along their beak and legs to offset low air temperatures.

▶ FEBRUARY 2 ◀

NICE MARMOT DAY

Groundhogs, a.k.a. woodchucks or *Marmota monax*, are some of the most recognizable rodents in North America thanks to their cameo appearances at annual winter celebrations, not to mention a title role in a Bill Murray movie. Legend has it, if the groundhog sees its shadow in the early hours of February 2, winter will linger for six more weeks. No shadow indicates spring is on the way. Meteorologically speaking,

FEBRUARY 2

World Wetlands Day marks the adoption of the Convention on Wetlands on this date in 1971. The annual event raises global awareness about the vital role wetlands play for people and the planet.

spring arrives March 1 (see March 21) no matter the groundhogs's prediction. In the wild, groundhogs are likely still in the throes of hibernation in early February, at least in Pennsylvania, home of Punxsutawney Phil. True hibernators, groundhogs remain in winter burrows from three months in southern states to upward of six months in the North.

While yellow-bellied and hoary marmots are found in the mountains of the West, woodchucks inhabit lowland meadows and fields across much of the East, throughout Canada, and extending into east-central Alaska. The name woodchuck is derived from the Algonquian word *wuchak* and not related to their dietary preferences. Groundhogs eat mostly vegetation but will opportunistically take insects. Not only accomplished diggers, groundhogs can also climb trees and swim.

▶ FEBRUARY 3 ◀

STAR LIGHT, STAR BRIGHT

The first star you see tonight will probably be relatively bright. It may even appear before the sky is perfectly dark. If it is twinkling, you're looking at a star not a planet. If you are in a big city, only the brightest of stars and planets can penetrate the light pollution scattering skyward. In International Dark-Sky certified locations (see April 1), like Capitol Reef, Death Valley, and Black Canyon of the Gunnison National Parks, there are countless stars, yet our eyes are drawn to the most illuminated ones. These are the reference dots people have connected into constellations for centuries.

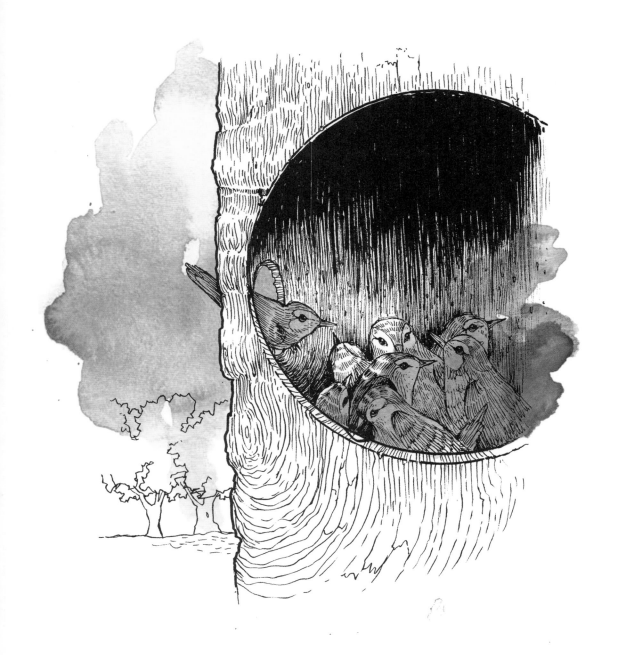

Communal Roosting, Pygmy Nuthatches

Even the faintest of stars shimmer. Brightness is a function of the distance the object is from Earth; therefore, astronomers differentiate between apparent and absolute magnitude. The sun outshines all the other stars because of its close proximity, not because it's the absolute brightest. Larger stars appear dimmer due to their distance from Earth. Stars are categorized on a logarithmic brightness scale. A magnitude 1 star is 100 times brighter than a magnitude 6 star. Lower values mean more dazzling stars. The apparent magnitude of the sun is -27. The North Star is a 1.97 apparent magnitude. Castor and Pollux, the iconic stars of the constellation Gemini, have magnitudes of 1.58 and 1.14 respectively.

> **FEBRUARY 4**

CUDDLE BUDDIES

As the temperatures plummet, animals must keep warm to survive. Many species of birds grow additional feathers for the wintertime, and mammals put on thicker coats of hair. Behavioral responses can also aid an individual's survival during those long winter nights. Sharing a sleeping space elevates temperatures and can keep birds from becoming bird-sicles. Nuthatches, winter wrens, and bluebirds may buddy up for the night, often with family units. One observer saw one hundred pygmy nuthatches enter the same cavity. Birds roosting as individuals also kick the thermostat up a notch or two by roosting in tree cavities. Live trees retain more heat overnight, but snags and nest boxes are also warmer than the ambient temperature. Mammals get in on the communal cavity roosting as well: a dozen or more flying squirrels have been known to pile into a single nest box.

> **FEBRUARY 5**

GOOD SAYS IT ALL

When Jane Goodall set up camp in Tanzania in 1960, she did more than document chimpanzee behavior in what's now Gombe Stream National Park; she changed the definition of what it meant to be human. Her work demonstrated that, in addition to chimps being omnivores (not vegetarians), the primates use twigs as tools to capture termites. At the time, the use of tools was what distinguished humans from other animals. She went on to discover many more parallel qualities between chimps and humans, including how chimpanzees experience adolescence, express affection, and even go to war. Goodall's first major publication, *In the Shadow of Man*, released in 1971, depicts her fieldwork in great detail.

In 1977, she founded the Jane Goodall Institute and later cofounded the youth organization Roots & Shoots. Both continue to propel conservation efforts worldwide while fostering a new generation of environmental leaders.

X MARKS THE SPOT

Even though wood frogs usually start calling first, one of the most iconic signs of spring in the East is the first peep of a spring peeper. These tiny frogs, about an inch long, let out piercing high-pitched peeps to ring in the warming temperatures. The first peeper calls generally range from November in southern states to March or April across New England and southern Canada. Temperatures of 40 degrees Fahrenheit are plenty warm for peepers to pipe up. Peepers, along with wood frogs, gray tree frogs, and bullfrogs, appear to be calling ten to thirteen days earlier in the year compared to records from the early 1900s. (Green frogs and American toads from the same region in central New York haven't shifted the timing of their calling.) Wood frogs and peepers are capable of freezing and thawing (see January 6).

Peepers come in an array of earth-toned colors, but all show a noticeable X across their backs. Their scientific name, *Pseudacris crucifer*, references this identifying cross mark. Members of the tree frog family, peepers have large sticky toe pads, although they spend much of their time on the forest floor. Spring peepers are occasionally fall peepers as they sometimes call after warm rains late into the year.

Spring Peeper

BAYOU THINGS

The flooded, murky backwater swamps of the Southeast are termed bayous. Thoughts of bayous tend to conjure up images of tea-colored waters, with gators lurking about in the shadows and wood storks perching in treetops. The term is often associated with Cajun and Creole cultures, but the word likely originated from the Choctaw word *bayuk* or "small stream." Bayous can be freshwater, saltwater, or brackish (a combination of both) and are teeming with wildlife, from invertebrates on up the food chain. Bayou Bartholomew stretches approximately 375 miles long from Arkansas to Louisiana and is considered the longest bayou in the world. More than one hundred species of fish alone have been identified here.

Bayou Sauvage National Wildlife Refuge is located entirely within the city limits of New Orleans, notably within the hurricane protection zone. Refuge managers use a series of levees, pumps, and flap gates to regulate water levels, mimicking seasonal conditions for plants and animals. The refuge

Swamp

offers critical habitat for wintering waterfowl, especially gadwall as well as American wigeon, northern shoveler, mallard, blue-winged teal, green-winged teal, and northern pintail. Summer-breeding ducks in the refuge include wood, mottled, and black-bellied whistling ducks.

FEBRUARY 8

WITCH HAZEL

American, or eastern, witch hazel is a late-fall bloomer, but February is prime time to see Ozark, or vernal, witch hazel shine. Both bring splashes of color in what can otherwise be a bleak-looking landscape. Witch hazel flower petals dangle down like ribbon, looking like octopus tentacles. As eastern witch hazel sheds its leaves, the yellow to orange blossoms burst out. This species gives off a spicy, rich aroma, adding to its appeal. The Ozark variety flowers before its leaves emerge in the spring, bringing a vibrant red-orange hue in contrast to the browns of winter. Both native species thrive as understory plants in deciduous forests—eastern witch hazel from Maine to Minnesota and south to the Gulf Coast; Ozark in the East and along the Ozark Plateau. Japanese and Chinese witch hazel, along with many hybrids, are popular ornamental plants.

The name witch hazel may be a reference to the early days of dousing for water with divining rods. The sticks would react to the presence of water, and the translations of *wicke* (lively) or *wych* (bend) became *witch* for the witching sticks.

FEBRUARY 9

IN THE YARDING

An animal's habitat includes the main things it needs to survive: food, water, shelter, and space. Food and water are easy to understand. The importance of shelter and space can be less obvious as an animal's needs may shift throughout the year. When the cold drags on, deer sometimes gather in large congregations. Known as yarding, this behavior brings herds together in dense thickets of conifers. Stands of conifers protect wildlife from winds and can maintain warmer temperatures than open areas. Research in northern Maine shows that stands of cedar, spruce, fir, and hemlock collect the most snow in the canopy. These woodlots can reduce snow depths at ground level by up to 40 percent compared to hardwood stands. Lower snow levels, along with many deer packing down trails, help the animals save energy when they move about. As any backcountry skier or snowshoer will tell you, it's a lot easier to follow a path than break new trail. While conifers provide shelter in late winter, deciduous trees remain an important browse for foraging deer.

PLAYING WITH FIRE FOR A CAUSE

Fire plays an important role in many ecosystems. Under the right conditions, prescribed burns can recycle nutrients back into the soil, promote vegetation growth, remove unwanted species, minimize pests and disease, and reduce fuel loads in an area. These habitat improvements can play a vital role in land management and can be beneficial for plants and wildlife. Controlled-burn plans factor in conditions, such as wind, humidity, temperature, and vegetation moisture. Fires aren't set unless strict criteria are met.

One species benefiting from strategic fire management is the red-cockaded woodpecker. This southeastern species is closely associated with the pines of the region. Historically, frequent fires cleared the understory without damaging the trees, and today, controlled burns replicate these conditions. Red-cockaded woodpeckers excavate cavities in live trees, especially ones infected with heart rot, a fungal disease that causes the wood to decay. The birds protect their cavities by tapping holes in the bark, leading to sap oozing, thus deterring tree-climbing snakes from raiding nests. The birds are somewhat cooperative within family units, as males will assist in raising future siblings by incubating, brooding, and feeding.

HIBERNATING BUTTERFLIES

As winter transitions to spring, be sure to keep your eyes peeled for mourning cloak butterflies. On the first warm days of the year, these brown-winged gems with festive yellow stripes become active. While it's not exactly hibernation, they do remain secluded all winter long in the northern states. Most insects survive the cold as eggs or larvae, but not mourning cloaks. They overwinter as adult butterflies nestled under the bark of trees. Mourning cloaks are some of the first insects active after the spring thaw. In warmer climates, mourning cloaks can be found any month of the year.

The state insect of Montana, mourning cloaks are widespread from Alaska to Venezuela and across Eurasia. In England, the species is referred to as the Camberwell beauty. Mourning cloak caterpillars feed on a number of deciduous tree leaves. Adults will feed on tree sap and can sometimes be enticed to backyard fruit feeding stations.

Moose

STARVING TO DEATH (OR NOT)

Moose are built for the cold—so much so that they can overheat even during winter. But staying warm is only one aspect of winter survival. They still have to eat. Or do they? For hibernators, it's not uncommon to go months without food, but for somewhat active ungulates (hooved mammals) like moose, finding food is a matter of life and death. It seems counterintuitive, but sometimes the best way to survive is to stop eating for a short period of time. Calories an animal takes in must remain greater than the calories they burn if they don't want to lose insulative fat. In the grips of winter, foraging can take high amounts of energy, and the quality of food available this time of year is also diminished. Instead of their summer diet of forty pounds of leafy vegetation a day, moose are stuck nibbling fir boughs or deciduous twigs. Moose spend much of the winter resting and ruminating (as in chewing their cud, not pondering their survival). This feeding cessation leads to a great decrease in body weight, and not all animals make it to spring.

CRUSTY STINK EYE

Backyard bird feeding is one of the most widespread nature activities. Just like you wouldn't serve your dinner guests a meal on a dirty plate, it's essential to keep your bird feeders clean and well maintained. Feeding stations can facilitate the transfer of any number of avian ailments, from salmonellosis to pox. Mycoplasmal conjunctivitis, also simply called house finch eye disease, was first observed in house finches in 1994 in the Washington, DC, area. The bacteria responsible for the pathogen had a long history in domestic poultry.

After a fairly rapid expansion across the continent, the prevalence of the disease in house finches seems to have stabilized at 5 to 10 percent of the population infected. Despite its common name, the disease infects a number of finches, including American goldfinch, purple finch, pine grosbeak, and evening grosbeak. Infected birds display swollen red eyes that can be runny or crusty. Long-term, the infection can impact vision and cause blindness. Regular cleaning and disinfecting of feeders will help minimize the risks for birds in your backyard. If you encounter sick individuals, take your feeders down for at least a few days, allowing the birds to disperse, and clean it thoroughly, thus helping to halt disease transmission.

LAKE EFFECT SNOW

In the Great Lakes region, every winter flurry brings out the armchair meteorologists spouting off about the lake effect snows. The thing is, though, not all snows result from this effect, even in the Great Lakes region. Conditions for lake effect snows occur when cold-air systems move over relatively warm open waters. The atmosphere takes up some warmth and added moisture, which leads to clouds capable of producing heavy bands of snowfall as the storm reaches land.

With prevailing winds from the west, locations on the eastern lakeshores are known for these heavy accumulation events—places like the Upper Peninsula of Michigan; Erie, Pennsylvania; Buffalo and Oswego, New York; and even Michigan City, Indiana. In rare instances, the winds can shift and dump true lake effect snows on western lakefront locations like Milwaukee, Wisconsin, or Toledo, Ohio, for example. But just because it's snowing by a lake doesn't mean this phenomenon is responsible.

PROJECT FEEDERWATCH AND THE GREAT BACKYARD BIRD COUNT

Project FeederWatch (www.feederwatch .org) morphed out of Long Point Bird Observatory's Bird Feeding Survey, which began in Ontario in 1976. Now the project covers all of North America and is coadministered by Bird Studies Canada and the Cornell Lab of Ornithology. From November to April, participants compile data from backyard feeding stations. This information helps researchers monitor the distribution and abundance of winter bird populations. Recording your data weekly is ideal for Project Feeder-Watch, but less frequent birding can also provide useful data. This program is great because it happens in your own backyard. And it is during the winter season, when birds are often easier to spot, but connecting with nature can take some extra encouragement.

FEBRUARY 15, 1564

Galileo Galilei is born. Known as the father of modern science, Galileo was a leading thinker in astronomy, physics, mathematics, and engineering.

The lead organizations, along with the National Audubon Society, put on a similar event during Presidents Day weekend each year. The Great Backyard Bird Count (http://gbbc.birdcount.org) includes participants from around the globe and feels like an online bird festival with near-instant updates and photos from the field.

MANGROVES

The dangling prop roots of red mangroves ring parts of coastal Florida, Puerto Rico, and the Virgin Islands. This tangle is prime nursery habitat for fish, crustaceans, and shellfish. The structures also act as sediment traps and help to accumulate organic matter and build up coastal soils. Found just upland and inland from red mangroves are black mangroves. Looking similar to cypress knees, this species is noted for pneumatophores extruding from the soil. These aboveground roots help the tree gather oxygen. White mangroves are farther inland and lack exposed roots. Thanks to adaptations that include the ability to block the absorption of salt in the roots or expel excess salt from the leaves, all mangrove species are highly salt tolerant. They grow at the transition zone linking coastal waters and inland habitats. These nearly impenetrable mangrove stands support rookeries of colonial nesting birds (see April 7). They are also beneficial to human inhabitants of the area. Mangrove forests serve as a buffer, reducing wave action, minimizing erosion, and absorbing floodwaters, especially during intense storms and coastal surges.

FIRE FALL

Depicted on the Yosemite National Park commemorative quarter, El Capitan is an epic granite wall that is the face of the famed California park. Legendary within rock climbing circles, El Cap is more than 3,000 feet tall, but you don't have to be a climber to reach the summit. The strenuous Yosemite Falls Trail to the top is best tackled in spring or fall. The panoramic overlooks from the top can't be beat, although in winter, the views can be even better from the valley floor below. Yosemite Falls spews over the edge and tumbles down nearly 2,500 feet in three separate cascades.

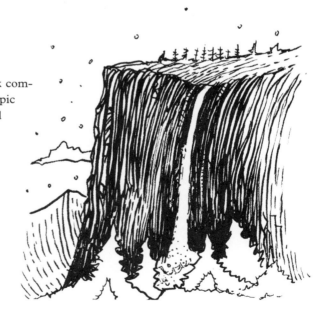

Horsetail Fall

A smaller but still impressive flow, Horsetail Fall draws added attention in mid to late February. This 1,000-foot-plus plume tumbles off the east side of El Capitan. Some evenings the setting sun casts an orange glow, creating a "fire fall" effect. (The glowing fall also is reminiscent of the park's former tradition of pushing burning embers over the edge of Glacier Point, creating a firefall on the granite.) If you can't make the trip to Yosemite, the National Park Service webcams stream many park features straight to your computer screen.

▶ FEBRUARY 18 ◀
SKY DANCER

When warming temperatures arrive in the eastern half of the continent, the skies come alive with the twittering flights of the American woodcock. As early as December in the South, and by March in the North, woodcocks perform their aerial sky dance as daylight turns to darkness. Along open meadows and fields near early successional forests and woodlands, these portly shorebirds gather. Males let out bold "peent" bleats from ground-level turfs. Suddenly they leap into flight, circling wider and wider as they fly high into the sky, eventually plummeting back to their original launch pads. Adding to the display, three modified and reduced outer wing feathers produce a fluttering whistle when the males are in flight. After breeding, females lay eggs and raise the young on their own, while males continue displaying for several more weeks.

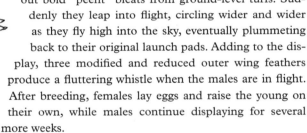

American Woodcock

American woodcocks have some of the greatest nicknames. Timberdoodle is the most common, but they also go by Labrador twister, bog sucker, night partridge, mud bat, brush snipe, and more. Many of these names reflect their feeding preferences as they probe the ground with flexible bills to munch on worms.

▶ FEBRUARY 19 ◀
QUILL PIGS

Despite a Latin name that translates to "quill pig," porcupines are actually large rodents. Never venturing far from trees, North American porcupines range across the boreal forests, western states, and south into northern Mexico. They remain active throughout the year but will hole up in a hollow tree during extreme weather. Their quills are stiff modified hairs that are quick to release, although they can't be projected as many people think. The tips are covered in microscopic barbs that help penetrate a perpetrator and make extraction a chore. Luckily for females of the species, young porcupettes are born with soft quills that don't harden until a few days after birth. Adults can have upward of thirty thousand quills and regrow new ones if they are ever lost to predators that venture too close. Agile fishers and mountain lions are some of the few known porcupine attackers.

▶ FEBRUARY 20 ◀
NOT MINIATURE FISH

While many people apply the name minnow to any small fish, an ichthyologist will tell you the term is reserved for the Cyprinidae family. This group includes the true minnows as well as carp (even the fifty-plus pound ones), barbs, and barbels. In terms of species, Cyprinidae is the largest and most diverse fish family. Extending to nearly six feet in length, the Colorado pikeminnow is the longest minnow species native to North America. The endangered fish inhabits the Colorado and Green River systems. Stocking efforts are also under way in the San Juan River Basin. The species is migratory, each year moving great distances to and from spawning sites.

Like other members of the family, pikeminnows lack teeth. Young feed on insects and plankton. When they reach a length of around four inches, their diets take a predatory turn. Adult Colorado pikeminnows feed extensively on fish, utilizing pharyngeal teeth in the back of the throat and along the gills to grasp prey. The fish was an important food source for John Wesley Powell's expedition down the Colorado River in 1869.

▶ FEBRUARY 21 ◀
ANTEATER TOADS

Have you ever thought you heard the long bleat of a lamb from the wild? Eastern narrow-mouthed toads start belting out their piercing lamb-like calls by mid-February in the southern part of their range, in the Florida Keys, for instance. The toads in the north, through the Mid-Atlantic and central Missouri (with an isolated pocket in southeast Iowa), don't get started until as late as April or May, although breeding can extend into October. Western narrow-mouthed toads have a buzzier, vaguely beelike call that sounds nothing like their eastern counterparts. This difference prevents hybridization where the two similar species overlap.

Microhylidae frogs, including both eastern and western narrow-mouthed toads, have pointed heads and plump round bodies. As the name toad implies, these species are fairly terrestrial. They have long toes on unwebbed feet. Narrow-mouthed toads specialize in a diet of ants, but they also eat termites, small beetles, and various invertebrates. Secretions from the skin help protect narrow-mouthed toads from predators and from ant bites.

AMONG THE WILDFLOWERS

Much like the fall colors of New England, the spring wildflower blooms of the Mojave Desert are iconic. Precipitation timing and amounts can greatly impact wildflowers. Some plant species will remain dormant for multiple years, but when the conditions are just right, they sprout and bloom in a burst

FEBRUARY 22, 1897

Ten national forests are established, including Bighorn National Forest in northern Wyoming.

of glory. Super-bloom conditions include well-spaced winter rains, followed by warm spring temperatures in the range of 70 to 80 degrees Fahrenheit and relatively calm winds that help maintain available moisture. This combination of factors comes together every ten years or so, and when they do, spring wildflower peepers descend upon the region. Exceptional locations for Mojave Desert wildflowers can be found at Anza-Borrego Desert State Park, Mojave National Preserve, and Death Valley and Joshua Tree National Parks. Lower elevations bloom earlier in the season, and high elevations peak a few weeks later. More than one hundred species add to this visual symphony, but desert lilies, lupine, verbena, primrose, and poppies are a few of the stars for this short-lived spring show.

Hedgehog Cactus Blossom

TOP-DOLLAR TRUFFLES

Culinarily, few foraged foods have the aura of the truffles. These subterranean fungal bodies fetch top dollar for their earthy umami. A couple European species dominate the dinner plates, but most species of truffle leave a lot to be desired when it comes to flavor. The Pacific Northwest is home to a number of truffle species often associated with Douglas-fir, mixed conifer, or oak stands, and a couple of species, including Oregon black and a pair of Oregon white types, are growing in popularity with commercial chefs and home cooks.

Truffles are also a favorite snack for flying squirrels, which may be the easiest way for you to find your first one. Look closely for disturbances on the surface of the soil that could be diggings where a critter has found a treat. Slight round mounds may also hint at a bulging truffle just beneath the surface. Raking for truffles is one way to harvest the bounty, but if you do, take care not to scar tree roots or upend too much soil and leaf litter. Hogs have traditionally been used as truffle finders. Dogs can also be trained to sniff out the goods. Truffles have a very narrow window of ripeness for harvesting and a short shelf life, so they are usually processed and preserved with oils, butters, or salts.

▶ FEBRUARY 24 ◀
FLYING THE FLYWAYS

Huge numbers of ducks can be seen in winter, but they start migrating northward earlier than you may think. By late winter, ducks are at their showiest in full breeding plumages. For many species, pair bonding is initiated and reinforced throughout spring migration. Many properties within the National Wildlife Refuge System serve a main purpose of providing habitat for waterfowl. The network of refuges connects wintering areas to breeding sites, with plenty of stopover rest areas along the way. Birds push north as weather conditions allow, often on the leading edge of melting ice. In winter, coastal areas and open water of the Great Lakes support huge rafts of diving ducks. San Francisco and Chesapeake Bays are great areas for watching sea ducks in winter. Flooded timbers of Arkansas are some of the most iconic wintering habitats for dabbling ducks.

Bird banding research from as far back as the 1930s has helped map the routes ducks, geese, and swans travel. Four flyway corridors have been identified: Pacific, Central, Mississippi, and Atlantic. These administrative zones help wildlife biologists manage migratory species more holistically.

▶ FEBRUARY 25 ◀
WHAT'S IN A NAME?

Thanks to a name change, the newest national park in the United States is now just a short train ride from Chicago. The area had been designated Indiana Dunes National Lakeshore since 1966, but in February 2019, Congress authorized the change to Indiana Dunes National Park. The same agency—the Department of the Interior—is still responsible for managing the area, so the change was partly symbolic. The christening may bring more attention to the region, but the landscape has always been spectacular. Nearby Indiana Dunes State Park is administered by the state and was established in

1925. The beaches are a big draw in the summer, but the natural landscape is impressive any season. In spring, ephemeral wildflowers dot the understory of the woodlands.

The sand dunes along the southern end of Lake Michigan tower nearly two hundred vertical feet above the water. The 3 Dune Challenge recognizes park visitors who conquer the tallest dunes, Mount Tom, Mount Holden, and Mount Jackson.

▶ FEBRUARY 26 ◀

SNOWPLOWING

Designated the mammal symbol of the United States in 2016, the bison is also the state mammal of Wyoming, Kansas, and Oklahoma. Bison are native to some of the snowiest places on the planet. Historically, millions of bison dotted the landscapes of the Great Plains from the Appalachian Mountains to Alaska. Both the plains and wood subspecies are built for extreme conditions. Thick hair provides such good insulation that snow doesn't even melt off from their body heat.

FEBRUARY 26, 1917

Acadia and Denali National Parks are established in Maine and Alaska, respectively. Two years later to the day, Grand Canyon National Park is created, and ten years after that Grand Teton joins the ranks.

To get to food, many animal species shovel away flakes by pawing, but not bison. Their large humps, broad thick shoulders, and oversized blocky heads function as snowplows, swiping drifts away from side to side. In the Badlands of South Dakota and the National Bison Range in Montana, edible grasses may be exposed in windswept areas. In Yellowstone, some bison take advantage of microclimates created by the geothermal areas in winter.

Bison

JOSHUA TREES

The Mojave Desert in Southern California is at its most vibrant when spring wildflowers are in bloom, but Joshua trees give this ecosystem a stunning complexity throughout the year. A type of yucca, Joshua trees typically grow at elevations between 1,300 and 5,900 feet. These hearty plants can survive temperature extremes and dry conditions, and yet a nonnative grass, urban pollution, and a changing climate are all impacting the species. Red brome, a Mediterranean grass, is spreading extensively, including to areas of Joshua Tree National Park. This species shift, along with an increase in vegetation growth driven by nitrogen pollution drifting in from coastal California cities, has led to more frequent and more intense wildfires in the park. Adding to concerns is that Joshua tree recruitment is down—young plants aren't becoming established.

The species has extensive but shallow root systems, which help the plants capture and store water during the desert's infrequent bouts of precipitation. Full-grown plants are fairly drought resistant, but young plants require more frequent rains. Climate models predict that with temperature changes of just a few degrees Fahrenheit, evaporation rates will increase, and the range of Joshua trees will become severely reduced.

TICK TOCK

In the southern states, ticks, especially black-legged (i.e., deer) ticks, can be expected throughout the year. Lone star and dog tick activity also picks up as the temperatures warm. Ticks are arachnids, not insects—go ahead and count the eight legs on the next one you find on yourself or your pet. After breeding, females lay thousands of eggs. Depending on environmental factors, these eggs can hatch within two weeks or remain viable for months. The larvae that hatch are six-legged and are generally referred to as seed ticks at this stage, regardless of species. After a blood meal, often from a small mammal, the tick will shed its skin, molting into the eight-legged nymph stage. Now poppy-seed-sized, the ticks seek out another meal of blood before developing into the adult stage. Adult ticks often find larger mammals as hosts, including ungulates (such as deer or hogs), dogs, or people, although birds, reptiles, and amphibians can also be susceptible.

Ticks can't jump or fly; instead they cling to the twigs of grasses or shrubs and wait for potential hosts to pass, a technique called questing. Ticks can detect body heat, moisture, breathing, and vibrations, and some species can recognize shadows. After being inadvertently picked up by a host, some ticks latch on right away. Others seek out thin-skinned segments of the body.

LEAP YEAR

While most years have 365 days, every fourth year gets a bonus day. Leap years are an attempt to recalibrate the calendar year with the astronomical year. It takes the earth 365.242199 mean solar days to make a lap around the sun. The extra minutes get added back to the calendar in the form of a leap day. Without this regular compensation, the seasons would slowly shift on the calendar. After just one hundred years, the calendar would be nearly one month off.

The concept of a leap year dates back to ancient Egypt. Julius Caesar first formalized it in calendar form in 46 BC, and Pope Gregory XIII adjusted the idea for the Gregorian calendar in 1582. There are still slight discrepancies, but those won't need to be sorted out for a few thousand years.

GLIDING MAMMALS

Bats are the only true flying mammals (see June 24), but a flying squirrel gliding from tree to tree is certainly an airborne aviator. Flying squirrels have a flap of skin between their front and hind feet called a patagium. These membranes provide enough lift, like on a fixed-wing aircraft, to allow the animals to glide between trees. They can travel more than one hundred feet in a single glide, and using their tail as a rudder, they have impressive maneuverability. They can't always ditch predators, though, and owls especially prey on these nocturnal rodents. Active all winter, flying squirrels will communally roost in a tree cavity for added warmth. They feed on a variety of things from seeds, nuts, and mushrooms to insects and bird eggs. Flying squirrels will take full advantage of a bird feeder and seem especially fond of peanuts.

MARCH 1, 1872

President Ulysses S. Grant signs a bill declaring the nation's first national park, Yellowstone, in Wyoming, Montana, and Idaho. The park now welcomes around 4 million visitors annually.

There are three species of flying squirrels in the United States and Canada. Northerns predominately live in the North, although their range extends along the Appalachian Mountains into North Carolina. Humboldt's flying squirrel, formerly classified as a subspecies of the northern, ranges from British Columbia to Southern California. Southern flying squirrels are the main species in the East and Midwest.

COMETS

Comets are frozen chunks of ice, dust, rock, or other metallic bits. The nucleus, or core, can be made up of ices formed from water, methane, ammonia, or carbon dioxide. Comets often originate from the Kuiper Belt or the Oort cloud in the outer reaches of our solar system. Many show an elliptical orbit of the sun, with perihelion being the closest point to and aphelion the most distant point from the solar body.

Multiple comets are visible each year, but proper viewing of many requires high-powered telescopes. The truly spectacular comets visible to the naked eye are few and far between. Halley's Comet, visible from Earth every seventy-five years or so, is perhaps the most famous. If you missed it in 1986, look to the skies around 2061. Hale-Bopp made an impressive showing in the late 1990s and remained visible for eighteen months, by far the longest known duration a single comet has been visible.

► MARCH 3 ◄

TRUE TORTOISES

The United States is home to five species of tortoises, all in the *Gopherus* genus and all living in the southern tier of states. The gopher tortoise is endemic (see October 8) to the sandy soils and longleaf pine woodlands and dry oak sandhills of the coastal southeastern states from Louisiana to South Carolina. It's recognized as the official tortoise of Florida and the state reptile of Georgia. The Lone Star State is home to the Texas tortoise. The Sonoran Desert tortoise ranges throughout much of western Arizona, while the Mojave Desert tortoise replaces it to the west. And then there is the Agassiz's desert tortoise, which was once classified as a subspecies but is now identified as a separate species inhabiting northwest Arizona and adjacent Nevada and Utah. Desert tortoises are the state reptiles of California and Nevada.

MARCH 3, 1891

Shoshone National Forest, the nation's first national forest, is established in northwestern Wyoming.

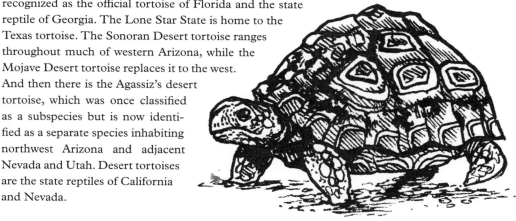

Desert Tortoise

Populations of all tortoise species are declining, and many are listed as threatened or endangered. Tortoises are long-lived but slow to reproduce. Males have concave lower shells, allowing them to be the "big spoon" when they mount females for mating. These bowling balls with stubby elephant-like hind legs lumber through desert scrub habitats during spring and fall; however, they spend upward of 95 percent of their time underground. Adept diggers, tortoises will burrow out tunnels ten feet long to stay protected from the weather extremes.

MARCH 4
SCENT OF THE SOUTH

The magnolia family is well represented across the planet by about one hundred species. Eight species of magnolia, and the closely related tulip tree, are native to the United States. Numerous other species and hybrid cultivars are used as ornamental landscaping trees, and these varieties tend to bloom earlier in the spring, often before their leaves emerge. Native species are especially prevalent along the Gulf and East Coast states. Magnolias are an old lineage of trees that relies on beetles for pollination. The stiff, leathery leaves appear floppy. And the flowers are made up of tepals, where petals and sepals are not differentiated.

Magnolias are noted for strong, rich aromas, and for some people, the scent defines the South. The southern magnolia is the state flower of Mississippi, the Magnolia State, though it was also declared the state flower of neighboring Louisiana fifty-two years earlier.

MARCH 5
SUGARBUSH

By late February or early March, sugar maple trees respond to the first signs of spring. After being dormant through winter, the tree sap begins flowing from the roots when the daytime temperatures creep above freezing. Spiles—small wooden or metal spouts—are tapped a couple of inches deep into the xylem layer of the tree. Larger trees can support multiple taps at once, and healthy trees can be re-tapped season after season. The process remains the same for commercial operations and backyard hobbyists. Sap is collected and converted to syrup in a boil-down process; it takes around forty gallons of sap to make a single gallon of the good stuff. Canada, especially Quebec, is the world's top maple syrup producer. In the United States, maple tappers are found in the upper Midwest and throughout New England. Sugar maple is the preferred choice, but syrup can also be made from other maples, box elder, sycamore, hickory, birch, and other tree species.

European Starling

SHAKESPEARE IN CENTRAL PARK

European starlings are unwelcome at most backyard feeders these days, but they had a fan in Eugene Schieffelin, a pharmaceutical manufacturer in New York in the nineteenth century. Along with other members of the American Acclimatization Society, Schieffelin promoted the exchange of plants and animals from one part of the world to another. He was especially interested in bringing all of the birds mentioned in Shakespeare's plays to North America. Scientists now understand, of course, that such relocation efforts have had major impacts on the ecosystems and have grave consequences for native species.

From an initial release of sixty birds on this day in 1890 in Central Park in New York City, with a follow-up of forty more the next year, the starling population in North America now numbers around 200 million. From backyard bullying for feeders and nesting sites to large-scale crop damage, the impacts of starlings are truly a Shakespearean tragedy.

YOU ARE GETTING SLEEPY

Hibernation is one of the most oversimplified concepts in nature. Many people think that hibernating animals simply sleep through the winter. As glorious as that sounds, there is much more going on, and researchers are still unraveling many of the mysteries. Hibernation isn't the same as taking an exceedingly long nap. Instead, body functions, including breathing and heart rate, slow down to an extreme, often operating at just 1 to 2 percent of average levels.

Some species, like marmots and many bats, are obligate hibernators, meaning they hibernate no matter the conditions and remain in that state for many months. Other species experience a facultative hibernation, meaning they can enter this physiological state as the environment requires, like when they're faced with extremely cold temperatures and limited food resources. Black-tailed prairie dogs will hunker down and hibernate for stretches at a time, but on those sunny days of late winter, they reemerge. Animals can stir from hibernation but will often wait until spring to crap out the anal plug that's been blocking the rear exit for the winter season.

Torpor is a more temporary, even just overnight, condition. Think of it as short-term hibernation. One of the most fascinating aspects of these decreased metabolic states is that animals don't experience the muscle loss that a person with a broken limb does, for example.

YOU GET A LINE, I'LL GET A POLE

Crawdads, crawfish, crayfish, mudbugs. These freshwater crustaceans have so many names, but did you know they aren't all the same critter? There are around 350 different kinds found in North America, and many have only scientific names. Crayfish play an important role in freshwater ecosystems, where they can alter microhabitats with their burrows and chimneys. Researchers estimate that upward of 20 percent are vulnerable to extinction due to threats like pollution and habitat loss. Tennessee alone has at least seventy-six different species of freshwater crustaceans. While some, like the various species of cave crayfish and the burrowing crayfish, are found only in isolated pockets of habitats, others are more widespread. Rusty crayfish are native to the Ohio River drainage but have become established across a wider range, turning invasive as they displace other local species. The signal crayfish of the Pacific Northwest can grow to nearly eight inches and one-third pound.

Recreational fishing for wild free-range mudbugs can pass the time, but to maximize efficiency, people typically use traps when fishing a crawdad hole. Check your local regulations, but these invertebrates can sometimes be harvested for human consumption or to be used as bait. There are a number of farm-raised crawfish operations too.

AN UNSILENT VOICE

Born in 1907, Rachel Carson was a published nature-loving writer by the age of ten and went on to become the voice for environmental protection. In college, she studied marine biology, later authoring numerous articles and publications regarding ocean studies. The book that launched her fame, *The Sea Around Us*, remained on the *New York Times*'s best-selling list for more than eighty weeks and won the National Book Award for nonfiction. While this book poetically described scientific findings from under the sea, Carson's marine research led her to a less glamorous subject matter. With the introduction of DDT (dichlorodiphenyltrichloroethane) in 1945, she began documenting the effects of chemical pesticides on marine wildlife and later on land habitats as well.

After many years of research, she bravely started writing the book she knew would spark controversy. *Silent Spring*, published in 1962, profiles real-life examples of how DDT damages wildlife, pets, agriculture animals, and humans. Carson was vilified by the chemical pesticide industry. But her words inspired a modern environmental movement, ultimately leading to the formation of the US Environmental Protection Agency and the banning of DDT. She died of breast cancer two years after the book was published.

Lava Flow

GO WITH THE LAVA FLOW

Volcanoes are one of the forces in nature that can fundamentally alter the landscape . . . instantly. These gaps in the earth's crust release molten rock, gas, or other debris to our planet's surface in a one-time event or continually over years or decades. Although other locations in North America have these geologic features, the Pacific coast states have notable concentrations of volcanoes. The Cascades are a chain of volcanic peaks stretching from Northern California to southern British Columbia. Alaska has plenty of volcanoes, but the real hotbed for volcanic action is Hawai'i.

Eruptions from a hot spot deep beneath the ocean surface along the Pacific Plate built up over time, eventually creating this chain of islands. The Big Island of Hawai'i is home to five volcanoes, including Kīlauea, one of the most active on the planet. Hawai'i Volcanoes National Park gives visitors a close encounter with the ropy flows of pāhoehoe (pronounced "paw-hoey-hoey") lava and the stony, rough, clinkery 'a 'ā ("ah-ah") form.

In northeast Wyoming, Devils Tower National Monument highlights phonolite porphyry, a type of magma that cooled and crystallized into hexagonal columns of rock. At New Mexico's Capulin Volcano National Monument, visitors can hike the crater rim and into the vent of an extinct volcano.

WHOOP IT UP

North America's tallest bird is also one of the most endangered. Whooping cranes stand erect at approximately five feet. These stately white birds were in serious trouble by the early 1900s. In the 1940s fewer than twenty remained. The species was included on the earliest versions of the Endangered Species Act. Historically they bred across the north-central United States and central Canadian provinces. A nonmigratory population was also found in southern Louisiana. The core population continues to travel between Wood Buffalo National Park in Saskatchewan and Aransas National Wildlife Refuge (NWR) on the Texas coast.

Creative conservation strategies have been implemented to aid recovery for the cranes. Captive breeding efforts help bolster wild populations, and caregivers dress in crane outfits to keep birds from imprinting on people. Sometimes ultralight aircraft is used to train whooping cranes to migrate. However, these efforts have delivered mixed results. An eastern migratory population now travels between Necedah NWR in Wisconsin and Florida's Chassahowitzka NWR and Alabama's Wheeler NWR. Nonmigratory flocks have also been reestablished in Florida and Louisiana. The whooping crane population is now estimated to be between five hundred and seven hundred individuals. With continued support, these icons can become a true conservation success story.

JOURNEY NORTH

The efforts of Journey North (www.journeynorth.org) are targeted at marking the arrival of spring and fall. More than fifty thousand sightings per year help map the seasonal movements of everything from birds and butterflies to worms and whales. Perhaps you've seen the migration maps the organization puts together each year. These visuals aren't possible without citizen scientists reporting what they see in their own neighborhoods. The anticipation builds as the wave of hummingbirds pushes north each spring. Monarchs are another focal species for Journey North. The group oversees a tulip project too. Bulbs are planted in test plots in the fall and then documented as they emerge and grow in the spring.

MARCH 13

BIG BABIES

After spending the winter in the relatively warm waters of Baja California, Mexico, gray whales begin migrating northward in March. Males and females without young are the first to depart Mexico, moving as far north as the Bering and Chukchi Seas for summer. As these giant mammals move north in the spring, they proceed along the West Coast of North America. Southern movements in the fall tend to occur farther out to sea. A roundtrip migration can be more than twelve thousand miles, making it one of the most spectacular treks in nature.

A gray whale's gestation period lasts longer than a year, so breeding occurs in the warm Mexican waters. Whales then move north before returning in the fall. Females give birth the winter following breeding. Adults are about as long as a school bus, and at birth, whale calves can be fifteen feet long and weigh fifteen hundred pounds. Nursing whales can consume more than fifty gallons of milk each day, and initially calves gain between fifty and one hundred pounds every day.

MARCH 14

STOCKPILING SNACKS

Similar to mockingbirds in appearance but far feistier, shrikes are carnivorous songbirds. Both northern and loggerhead shrikes are known for their food-caching technique. These not-quite-robin-sized predators feed on anything from large insects to small mammals. They also target birds, reptiles, and amphibians. Shrikes often dispatch prey species with a bite to the nape, and sometimes they store these food resources by impaling them on plant thorns and barbed wire or by wedging them in the forks of tree branches. Storage is especially useful during the breeding season and in winter, and these food caches may help males attract mates. Plus, delaying the consumption of some poisonous prey species, like monarch butterflies and eastern narrow-mouthed toads, allows the toxins to break down. You may sometimes

MARCH 14, 1903

President Theodore Roosevelt proclaims the nation's first national wildlife refuge, Pelican Island, in Florida in the Indian River Lagoon on the Atlantic coast.

hear shrikes referred to as butcher-birds, and although the North American shrikes aren't in the same family as the butcher-birds, both groups share similar feeding behaviors.

▶ MARCH 15 ◀
TREE BEARDS

Spanish moss, neither a moss, nor from Spain, is an epiphyte in the bromeliad family. With a core range in Central America, the plant is native from central South America north to Virginia. French explorers called it Spanish beard because the plant reminded them of the facial hair of the conquistadors. Grayish-green Spanish moss hangs in long festoons from tree trunks and branches, capturing nutrients and moisture from the air with specialized scalelike filaments. Broken-off strands can regenerate, but the plant also releases seeds that drift on the wind before settling and sprouting. It grows especially well on oak and cypress trees.

Spanish Moss

Spanish moss has been used for centuries for everything from livestock feed to mattress stuffing. Thanks to its water-retention capabilities, it can often be found in floral arrangements. Birds use Spanish moss for nesting material or roost in the clumps, as do bats, snakes, lizards, and frogs. One species of jumping spider has been documented exclusively in Spanish moss.

▶ MARCH 16 ◀
CRANE CONGREGATION

While nonmigratory populations of sandhill cranes are already well into the breeding season in the southeastern United States, mid-March is the peak of migration for cranes moving through central Nebraska. Each March the majority of the world's sandhill cranes, some 600,000 individuals, congregates along a sixty-mile stretch of the Platte River as they move between winter and summer homes. The braided channels of the wide but shallow river provide overnight roosts for the lanky slate-gray birds. During the day, the omnivorous birds feed in the surrounding fields and wetlands, eating up cultivated grains, seeds, tubers, grubs, worms, snails, and even small rodents, amphibians, and reptiles.

Each bird will spend upward of two weeks at this critical stopover habitat before moving onward, riding warm thermal airwaves to help carry them north to Canada, Alaska, and far eastern Siberia. Each morning up and down the Platte River between Kearney and Grand Island, a cacophony of crane calls overwhelms the senses as flocks take off at first light. Courtship dances and pair bonding occur along the entire migration route, adding to the fascination of this seasonal journey.

▶ MARCH 17 ◀
NOT EASY BEING GREEN . . . FINNED

Originating 150 million years ago, bowfins, often known as dogfish, have persisted since the time of the dinosaurs. As the breeding season heats up, male bowfins take on a vibrant shade of green along their pectoral, pelvic, anal, and lower caudal fins. Unlike the other primitive fishes (e.g., gar, sturgeon, paddlefish, and polypterids), bowfin males actively invest in parenting. They guard their eggs and protect young bowfins until they reach a couple of inches in length.

Bowfins have elongated dorsal fins that aid in the ability to swim backward. Bowfins are able to tolerate low levels of oxygen in the water since they can breathe air from a modified gas bladder. They also have tubular nostrils that stick out from the head, giving them an awkward appearance but a keen sense of smell. At best bowfins are often disregarded, and at worst the species is actively harvested and discarded wantonly based on the false notion that they decimate fish populations. Habitat loss, especially decreased aquatic vegetation, negatively impacts these ancient survivors.

▶ MARCH 18 ◀
WHITE AS A GHOST CRAB

After spending the winter in burrows up to four feet underground, ghost crabs resurface as the temperatures rise. Somewhat nocturnal and sensitive to human disturbance, ghost crabs are a sometimes-overlooked beach critter that lives about three years on average. Armed with flashlights, people can catch a quick glimpse of them scurrying about after hours. You may also spot these pale sand-colored crustaceans beachcombing for a scavenged meal brought in by the ocean surf. These two-inch crabs have tiny pincers used for foraging and for posturing as males try to compete for mates. Ghost crab eggs are sent out to sea, where larvae hatch and develop.

Ghost Crab

Their eye stalks give crabs a 360-degree view of their surroundings but leave them vulnerable to a blind spot directly above. Common crab predators include raccoons and gulls. *Ocypode*, their genus name, means swift-footed, and these terrestrial crabs have been clocked at ten miles per hour—a most-impressive crab walk.

MARCH 19

THE GLUTTONOUS GLUTTON

Given that its scientific name, *Gulo gulo*, means "gluttonous glutton," it should come as no surprise that wolverines have hearty appetites. Circumpolar in distribution, these large weasels are found throughout the northern mountains, forests, and tundra. Wolverines are opportunistic feeders. They often scavenge dead carcasses and seem to key in on avalanche chutes in the spring. Apt predators, they can take down medium-to-large mammalian prey—from snowshoe hares, porcupines, and marmots to the occasional deer and caribou—especially in deeper snow conditions.

Delayed implantation, where the embryo remains unattached to the uterine wall for a period of time, allows females to give birth to snowy white kits in winter dens. Wolverines remain active all year, and adults cover extensive home ranges during their annual cycles, scent marking as they go. Female territories sprawl one hundred to two hundred square miles, and males can double that.

MARCH 20

MIGRATION RESTLESSNESS

Many migratory birds experience a period of fidgetiness in the springtime. Called *zugunruhe*, German for "movement or migration restlessness," the behavior was first noticed centuries ago in captive birds. During migration season, these animals jumped, hopped, and fluttered about. This movement wasn't random either, as it lined up with the direction the species' natural migration would take them. Later, researchers found that altering the magnetic fields would cause the captive birds to adjust their direction. Birds likely can't differentiate between north and south based on these magnetic field clues; instead they get a general sense of latitude because the force is stronger at the poles than at the equator.

The sun and the stars aid in navigation for at least some species of migrators. Day length (photoperiod) also plays a role in triggering movements, as does ghrelin, a hormone that controls hunger. Birds utilize stopover habits during their long journeys. They refuel and build up fat reserves before continuing, so the link between ghrelin and migration isn't as far-fetched as it may seem. There is clearly more to migration than birds flying south for the winter and returning north in the spring.

SPRING

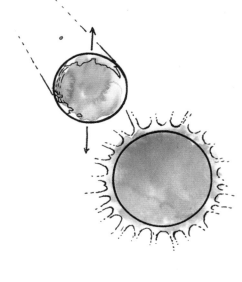

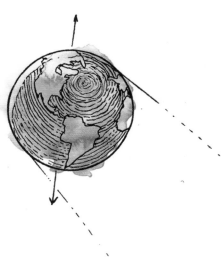

SPRING EQUINOX

SPRING EQUINOX

For naturalists, spring is the season of new begin-
nings. It kicks off the year. Meteorologically, the seasons are divided quarterly, with spring commencing
on March 1 and running until the end of May. Astro-
nomically, though, March 21 is the Spring Equinox
for the Northern Hemisphere. Here's why: The earth
is tilted 23.5 degrees, and the North Pole is always
pointing toward the North Star, which means the
Northern Hemisphere points toward the sun during
part of its orbit (summer) and away from the sun on
the other side of the orbit (winter). The Spring and
Fall Equinoxes occur between those points of the orbit.

MARCH 21, 1933

President Franklin Delano Roosevelt
proposes the Civilian Conservation
Corps (CCC) as a major component
of the New Deal. The bill is drafted
and signed into law just two weeks
later on April 5. The CCC crews
were responsible for construction
projects and habitat work, especially
at many national parks and forests.

Both the Spring and the Fall Equinoxes represent when the sun rises and sets directly over the
equator, which makes for roughly equal days and equal nights, although atmospheric conditions
alter this slightly. It is also a bit uneven because sunrise is when the first rays of the sun are detect-
able, while sunset happens when the last part of the sun dips below the horizon.

I SMELL A SKUNK . . . CABBAGE

Skunk cabbage is one of the first plants to sprout each year. It starts growing in late winter, especially
in damp woodland areas of New England, the upper Midwest, and south through Tennessee and
North Carolina. Sometimes the red-green-and-yellow-patterned modified leaves, called the spathe,
poke up like a cone through the snow. Huge green leaves grow in as temperatures warm in the spring.
Skunk cabbage is unique because the plant gives off a slight bit of heat, creating a microclimate that
melts away nearby snow.

Named skunk cabbage for a reason, the plant emits a foul stench that keeps many animals from
eating the cabbage-like leaves. The odor also attracts flies that act as plant pollinators.

Curiously, there is also a skunk cabbage in the Pacific Northwest and Alaska. This related plant
shares the characteristic odor and gigantic leaves, but it's a yellow spring bloomer that doesn't give
off heat like the eastern variety.

LITTLE STINKERS

Speaking of skunks . . . their stench has been immortalized by the lyrics of the 1972 hit song "Dead Skunk," by Loudon Wainwright III: "You can feel it in your olfactory." It's true: the spray of a skunk packs a punch. But it isn't like the animals are out spouting off all the time. Boldly marked, the skunk's black-and-white coloration is meant as a warning flag. Skunks will posture in an attempt to ward off any would-be agitators, pawing their feet, hissing, even bluff spraying. Finally, as a last resort, skunks spray.

A skunk can fire off a squirt at least ten feet away from its body—about the length of four or five small pet dogs, for example. A sulfur compound, the potent juice is energetically costly for skunks to produce. It takes about ten days to restock the spray supplies in the anal glands. The scent will probably linger on your pet for even longer. Tomato baths are one suggestion for washing away the odor, although a shampoo of hydrogen peroxide, baking soda, and dish soap is likely more effective.

Striped skunk is the most widespread species, followed by western and eastern spotted skunks. Hooded and hog-nosed skunks are found in Mexico with ranges extending into Arizona, New Mexico, and Texas.

Spotted Skunk

MIGHTY MORELS

Morels

More northerly locations will have to wait a few more weeks, but morel mushroom season kicks off in the southern states by the end of March. Some old-timers claim that when the mayapples are up, the redbuds are in bloom, or the lilacs blossom, that's when morel mushrooms are ready to pop too. Soil temperatures need to reach approximately 45 to 50 degrees Fahrenheit for fruiting bodies to emerge. There are a handful of morel species across the country (researchers continue to debate the taxonomy). Each type is found under different growing conditions. Some are associated with conifers. Others tend to grow near certain deciduous plants. The *Morchella esculenta* is fairly widespread in the East.

Not all mushrooms that look like morels are morels, though. False morels can look similar in appearance, although they have solid stems. If you plan on eating these delicacies of nature, be certain you have identified them correctly. With any edible mushrooms, it's always a good idea to try only a small bite the first time you eat it—reactions can very, even with widely edible varieties. The morel was declared the state mushroom in 1984 by Minnesota, the first state to make a mycological proclamation.

MIGRATING UP AND DOWN

It probably falls short of an official migration, but each year, worms make seasonal movements. Before the onset of winter, these invertebrates burrow a few feet down into the soil, working their way below the frost line. As ground temperatures rise in the spring, the worms tunnel upward. American robins certainly notice the return of the worm. Even if they overwinter in your area feeding on berries, the first robin scampering across the backyard eyeing a fresh worm morsel is a sure sign of spring.

Gardeners often embrace worms as much as robins do. Worm composting, or vermicomposting, creates nutrient-rich castings that enhance the growth of plants. Not all worms are good news for the soil or garden, though. In some places, nonnative worms are taking over (see June 7). Journey North (see March 12) asks citizen scientists to monitor the arrival of worms in their own backyards.

TEMPORARY POOLS

Vernal pools, also called ephemeral ponds, are temporary in nature and are of critical importance for many species. These seasonal wetlands hold water for an extended time frame and dry up by late summer. This scenario isn't viable for fish, but for this very reason, ephemeral ponds are valuable to a whole suite of invertebrates and amphibians. Fairy shrimp is one such invertebrate that thrives in these temporary wetlands. With translucent bodies, they can be tough to spot, but look for the half-inch-long bodies fluttering in the water column. Observe them early in the season, well before most predators are active in these ecosystems.

Many species of frogs and salamanders deposit eggs, hatch out, develop, and grow in these aquatic environments before undergoing metamorphosis and moving into upland habitats. Fish aren't a worry in these temporary wetlands for the amphibians, but mammalian and avian predators are still a threat. Another risk is that the waters dry up before the young are fully developed. These shallow wetlands are also particularly vulnerable to changing environmental conditions and to pollution and contamination.

Fairy Shrimp

FIDDLEHEADS

Fern shoots spring up as the days get longer. The name fiddlehead doesn't refer to a species of fern, instead the new growth is called that because the fresh, tight sprouts resemble violin scrolls. Widespread from Alaska to the Canadian Maritimes and south through the Midwest and New England, the ostrich fern is the species most often harvested for consumption.

It's recommended that all ferns be cooked before they are eaten. A quick sauté in butter with some garlic, shallot, or onion will showcase their fresh, earthy, grass notes. Note that not all ferns are edible, and a few can cause stomachaches. One way to extend fiddlehead season is to pickle the ferns for eating throughout the summer.

Fiddlehead

BUDBURST

Learning about botany involves numerous opportunities to engage in citizen science. One project even has two different ways to participate. Budburst (https://budburst.org) is administered by the Chicago Botanic Garden and has dozens of partnering organizations and agencies throughout the country. Volunteers track the phenological timing of individual plants over the course of a growing season, which works well for observation at a local park or in your own backyard. You follow the progress of each plant, watching for the subtle changes from day to day and week to week.

Another way to pitch in is to document the stage of a plant at any point. If you're traveling, for example, you can still let researchers know what stage of development the plants are in on that day. Budburst has various criteria to standardize sampling across five groups of plants, including conifers, deciduous trees and shrubs, evergreen trees and shrubs, wildflowers and herbs, and grasses.

EASTER FLOWER

Eastern pasqueflowers, sometimes referred to as prairie crocuses, are early bloomers in the grasslands of the central US and southern Canada. Pasqueflower is the state flower of South Dakota and the provincial pick for Manitoba. The feathery, silky stems and buds give the plant a hairlike appearance, and this fuzzy pubescence can keep leaves up to 10 degrees warmer than the surrounding air. Blooms are most often tinged a lovely purple. Later in the year, the seed heads look like a tuft of downy feathers.

END OF MARCH

Earth Hour is celebrated at 8:30 p.m. local time on a Saturday at the end of March. The global event, organized by the World Wide Fund for Nature and partnering organizations, encourages people to turn off nonessential electric lights as a symbolic commitment to the planet.

The common name references a bloom time that roughly correlates with Easter. The word pasqueflower is derived from *pasakh* the Hebrew word for Passover. As a whole, pasqueflowers are more diverse in Europe, with over 30 species native to that continent.

DANCE LIKE EVERYONE IS WATCHING

Each spring, many species of birds gather on traditional breeding grounds known as leks. At least ten families of birds use lekking for breeding worldwide, including many species of grouse in western North America. Dozens of males congregate in the early-morning hours to put on elaborate displays, seeking the attention of females. For sage grouse, sharp-tailed grouse, and prairie chickens, these lek displays involve inflated air sacs, strutting, dancing, and the occasional skirmish between rival males. Upward of 90 percent of females breed with the master cock of the lek, who holds the prime territory near the center of the lek. Postcopulation, females nest and raise the young, while males continue lingering on leks. Viewing opportunities, from roughly March to May, offer observers a chance to fully immerse in this natural wonder, often from the comforts of an automobile or a blind.

LITTLE TORCH

No joke, for much of the year ocotillo looks like clusters of dead sticks jutting out of the rocky soils of the Sonoran and Chihuahuan Deserts. The twenty-foot-tall shafts are covered with menacing spines. When enough moisture is present, short green leaves sprout along the stem. They dry up and are shed quickly, but leaves can reappear multiple times throughout the growing season.

Between March and June, the shrub transforms, looking especially alive as an explosion of flame-red tubular flowers blooms from the stem tips. The tube flowers are visited by hummingbird pollinators and are an important food source in dry landscapes from West Texas to Southern California. *Ocotillo* means "little torch" in Spanish, and the reason behind this name becomes obvious in the springtime. The plant is also sometimes called candlewood, flaming sword, slimwood, coachwhip, and vine cactus.

THE DARKEST NIGHTS

Light pollution outshines stars on a growing footprint of the planet. The International Dark-Sky Association aims to preserve and protect the nighttime environment through policy work and education about responsible lighting. A lot of areas are dedicated to making continual progress as best lighting practices are constantly improving. Based on criteria for darkness and a dedication to the preservation of the night sky, Flagstaff, Arizona, was declared the first International Dark-Sky Place in 2001. The National Park Service has a number of units certified as International Dark-Sky Parks, and some state and county parks also participate. These destinations host star parties and astronomy events throughout the year, but big efforts coincide with the new moon of April. This is when International Dark-Sky Week takes place. The event, first proposed by high school student Jennifer Barlow in 2003, is a celebration of the night.

APRIL 2

MILKY WAY

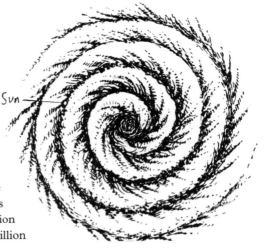

On the darkest nights, the Milky Way is most visible. Home sweet home for all of us earthlings, this galaxy is so big that it would take light about 100,000 years to cross it from side to side. Our planet is camped out in one of the arms of this spiral galaxy, known as a bar spiral, roughly halfway between the center and the edge. The middle of the galaxy is likely a supermassive black hole. The whirlpool-shaped Milky Way is always rotating, and each pivot takes 200 million years. Within the galaxy are approximately 100 billion stars, none more important to Earth than our sun.

The Milky Way is just one of forty galaxies in the so-called Local Group. The Andromeda Galaxy rivals our own for size, while the others are somewhat smaller. It's nothing to lose any sleep over, but in roughly 4 billion years, these two galaxies could collide.

APRIL 3

DANCE PARTY OF ONE

While the prairie grouse breed on communal gathering grounds known as leks, the males of the woodland species—including dusky, sooty, spruce, and ruffed grouse—use solitary displays to attract hens. Dusky, sooty, and spruce grouse inflate air sacs along the throat to emit a low hooting sound that penetrates the forested landscape. Ruffed grouse, named for a dark ruff of feathers visible on displaying males, strut their stuff back and forth along downed logs. They stop occasionally to rapidly flap cupped wings, creating whooshing sounds that resemble the putt-putt-putt of a tractor starting on a cold morning. Listen for displaying dusky grouse in the Pacific Northwest, sooty grouse in western mountains from New Mexico to northwest Canada, spruce grouse from boreal Alaska and Canada and south into central Idaho, and ruffed grouse in mixed woodlands from the Midwest and southern Appalachian Mountains through the north woods and into the central Rockies.

Dusky Grouse

▶ APRIL 4 ◀

A LEAGUE OF HIS OWN

The Izaak Walton League is an organization of conservationists across 240 community-based chapters who work to preserve the quality of soil, air, woods, waters, and wildlife. It started in 1922, when a group of sportsmen became fed up with the destruction of forests, wetlands, and wilderness that were threatened by the irresponsible use of resources at the time. The gentlemen decided to name their group after the seventeenth-century English biographer and author Izaak Walton—famous for the literary classic *The Compleat Angler* (1653).

The book is a poetic manual of sorts, detailing how to catch all different kinds of fish. But more than that, it explores what it means to find peace in nature, friendship, and country living. Three and a half centuries later, it's one of the most reprinted books in the English language.

▶ APRIL 5 ◀

WATER WEASELS

The most aquatic of the weasels are the otters, and two types live in North America: sea otters and river otters. Sea otters live in the Pacific Ocean and spend the majority of their lives in the water. They can give birth anytime during the year, but peaks occur in California in spring and fall, while Alaska sea otters give birth between May and June. The fully furred young sea otters are born in the water, although they remain dependent on their mothers for a few months.

River otter pups are born in dens or burrows. They weigh less than a pound at birth and are blind, toothless, and practically immobile. They spend the first few weeks of life nursing and sleeping in these maternity dens. River otters were wiped out of much of their historic range by the early 1900s, but by the beginning of the twenty-first century, extensive reintroduction efforts restored populations to twenty-two states.

CHERRY BLOSSOM SPECIAL

Each spring the Washington, DC, area comes together to celebrate the National Cherry Blossom Festival. The trees of honor are Yoshino cherries, and they were a gift to the city in 1912 by Tokyo mayor Yukio Ozaki. The States reciprocated with a gift of flowering dogwood trees to the people of Japan three years later. Today, the DC festival embraces these trees as a symbol of enduring friendship between the citizens of Japan and the United States.

The Cherry Blossom Festival runs from late March to early April during the peak for flowering buds. National Park Service horticulturists categorize the progression into five steps: green buds, florets visible, extension of floret, peduncle elongation, and puffy white. Peak is defined when 70 percent of the trees are showing full bloom. Yearly weather patterns affect the dates, but long-term analysis shows that since 1921, the dates for peak blooming of the cherry trees at Tidal Basin have shifted to approximately five days earlier. The festival has even missed peak a few years! The nation's capital isn't the only home to impressive cherry blossom displays. Other hot spots include Macon, Georgia; Portland and Salem, Oregon; Nashville, Tennessee; Brooklyn, New York; and Seattle, Washington.

APRIL 6

California Poppy Day, a celebration of the state's official flower, is in early April, while Poppy Week isn't until mid-May.

Cherry Blossom

COLONIAL WATERBIRDS

The so-called colonial waterbirds is a collection of species that tend to nest in large groups. The term can apply to gatherings of ground-nesting birds like gulls, terns, and skimmers or to shrub and tree nesters like herons, egrets, and ibis. These rookeries can be home to hundreds of nests from upward of a dozen different species. Within the congregation, birds can be territorial, each carving out their little rookery nook. Cumulatively, the group can be beneficial in providing protection from predation as there are more eyes to spot any would-be nest raiders. Limited suitable nesting sites may also force the highly concentrated arrangements. The skies around rookeries can look like an airport as birds cruise in and out, bringing food back to nest sites. Coastal habitats can provide prime rookery-viewing locations, but it's best to observe from a distance.

APRIL 8

SPOONBILL

Prehistoric in appearance, paddlefish have remained structurally similar since the Late Cretaceous Period 70 to 75 million years ago. Curiously, the fish's diagnostic paddle snout (rostrum) doesn't begin to form until after hatching. Adult paddlefish are torpedo shaped, smooth skinned, and scaleless fish with a deeply forked tail and a large spoonbill. American paddlefish are filter feeders and survive primarily on zooplankton. The rostrum is covered in specialized electroreceptor cells that aid in foraging by detecting the weak electrical fields of the plankton, which comes in handy since the fish have poorly developed eyes.

Native to the Mississippi River drainage, paddlefish were once present from Montana to New York and south to the Gulf Coast, but the species has been extirpated from many regions, and populations are continuing to decline. The species is rarely targeted by bait anglers; instead paddlefish are snagged or speared during short harvest seasons. They can grow to impressive sizes, often tipping the scales at well over fifty pounds. The flesh is edible, but equally as prized is the fish roe, which is cured and eaten like sturgeon caviar.

APRIL 9

SALAMANDERS OF THE APPALACHIANS

The tropics are often cited for their biological diversity, but for one group of amphibians, the Appalachian Mountains are the global hot spot. Though salamanders are well represented in North America, the southeastern mountains support the highest concentration of unique species. Called the salamander capital of the world, Great Smoky Mountains National Park is home to thirty species. The entire region supports seventy-two species. The cool, damp environs are ideal for amphibians, especially the species of lungless salamanders that breathe through their skin. Many species, like

Marbled Salamander

red-legged, Pigeon Mountain, and red-cheeked salamanders, have limited ranges and are endemic (see October 8) to a small geographic area, sometimes just a few drainages on a mountain. Others, including marbled and spotted, are more widespread with ranges stretching beyond Appalachia. Despite spending much of their lives unseen by humans, salamanders make up a substantial portion of the biomass in the region. They are sensitive to environmental conditions and are important indicator species.

The largest salamander in North America, the eastern hellbender, also has an extensive range, from New York to Georgia. Entirely aquatic, it is the state amphibian of Pennsylvania.

▶ APRIL 10 ◀

MYRMECOCHORY

Pollinators get a fair bit of attention, and rightfully so, but this spring, pause to appreciate ants and myrmecochory—the dispersal of seeds by ants—too. Early in the season, when there are few pollinators around, spring wildflowers often rely on self-pollination. Then ants take over. They are essential in moving the seeds of a number of plants. Perhaps as much as 5 percent of all flowering plant species worldwide rely on this method of seed dispersal. It is especially common in fynbos shrubland habitat of South Africa, but some species in the eastern deciduous forests of North America, including many spring ephemeral flowers, also rely on ants.

APRIL 10

Florida celebrates Gopher Tortoise Day on this day every year, as designated by the Gopher Tortoise Council, which works to "conserve the gopher tortoise and the fascinating world in which it lives" (see March 3).

The trilliums, hepatica, trout lily, violets, wild ginger, squirrel corn, and Dutchman's breeches are all myrmecochorous species. They have seeds with elaiosome coatings, fleshy structures attached to the seeds. The ants consume these nutrient-rich packets and the seeds are left to sprout.

▶ APRIL 11 ◀

OVER THE GLACIAL HILL

Long after a glacier retreats, piles of debris called moraines are left behind to tell its story. Terminal moraines form at the front end of a glacier. They mark the farthest point the mass of frozen ice and rubble extended. Lateral moraines flank the sides of a glacier. Moraines can stand hundreds of feet tall, and the outwash till of rocks can include boulders the size of houses. Glacial ponds are often left in the wake of the retreating ice, formed by chunks of ice that broke off from the main body and eventually melted. Many of the lakes or ponds in the upper Great Lakes region have origins that can

Moraine

be traced back to the Wisconsin glaciation ten thousand years ago. Mount Rainier in Washington state is the most heavily glaciated peak in the Lower 48.

Another great location for exploring glacial features is Grand Teton National Park in Wyoming. Glaciers carved many of the canyons within this mountain range. These broad U-shaped cuts were slowly scoured out. The glacial advance and retreat left behind hilly moraines and glacial lakes along the foot of the range, including Leigh, String, Jenny, Bradley, and Taggart Lakes and the Potholes. Glacier National Park in Montana is a great place to view glacial processes at work, although some scientists predict that these glaciers will disappear as soon as the year 2030.

APRIL 12
BLOODROOT

Another of the myrmecochorous species (see April 10) is bloodroot. Like the other early-season wildflowers of the forest floor, bloodroot has a short window for blooming. By summer, these locations are fully shaded, but dappled sunlight filters through the tree canopy early in the season before leaf out. A white beauty with bright-yellow stamens, the bloom is on a short stalk offset by a single large palmate leaf that unfurls slowly over the course of days as the plant blooms. The flowers close at night and reopen each day.

The name bloodroot is a reference to the orange liquid the plant exudes when broken, which can be used as a natural dye. In addition to dispersal by ants, bloodroot can spread via orange rhizome networks growing in the soil.

APRIL 13
CITIZEN SCIENCE DAY

Citizen science has always been a hallmark of natural history. Fields of study from astronomy to zoology have been shaped by people who weren't necessarily employed in related professions. The first widespread coordinated effort to engage amateurs in data collection was likely the Christmas Bird Count (see December 25) in the early 1900s. These days, there are thousands of ways to become involved. The Citizen Science Association and SciStarter grew out of this extensive network of research happenings. Also referred to as public participation in scientific research, the concept aims to connect volunteers with organized research. Now there is an annual conference and a peer-reviewed journal for these types of projects. Additionally, Citizen Science Day is celebrated in mid-April each year.

ICE-OUT

One indicator of the demise of winter, at least in the north country, is ice-out. Plenty of communities keep tabs on the arrival of spring by monitoring the first days of open water. Some make it a fundraiser by taking friendly bets on the exact minute the ice will melt, often measured by a visual cue, such as when a gutted car sinks, although fish habitat structures, discarded Christmas trees, and plenty of other things have been used. Monitoring ice-out is a safety precaution to keep anglers off of thin ice. It also allows researchers to assess climate trends.

Year to year, ice-out can vary by weeks in a given location. Southern Minnesota bodies of water free up in mid-March, while lakes in the north of the state remain frozen until mid-May. Ice-out for Lake Minnetonka (southwest of Minneapolis) has been recorded annually since 1855 with the median (half before and half after) date of April 14. The latest date for ice-out at Minnetonka is May 5 and was recorded in 1857 and again in 2018. Earliest dates include March 11 in 1878 and March 17 in 2016.

► APRIL 15 ◄

HOW SHREWD

Shrews are some of the smallest mammals on Earth, but they aren't rodents. Instead these tiny fur balls, about the size of a cotton ball, are more closely related to moles. Seemingly hyperactive, shrews scurry about in near constant feeding frenzies, fueling their high metabolisms. Their hearts can beat up to eight hundred times per minute. Shrews have pigmented incisors, and some species are slightly venomous, helping them to take down prey larger than themselves. Shrews may use a simple echolocation to maneuver about their habitat. The masked shrew, *Sorex cinereus*, is the most widespread of the North American shrews, while the heftier short-tailed shrew, *Blarina brevicauda*, is abundant in the East. Most close encounters of the shrew kind involve finding a dead one, although a few lucky naturalists have seen shrews scurrying about in the leaf litter, undoubtedly looking for their next meal. Look for one poaching seed from under a backyard bird feeder.

Masked Shrew

SHOW ME YOUR DANCE MOVES

Most birds, like American robins, sing to impress potential mates, but not the grebes. They rely on their dance moves. Western and Clark's grebes breed on open-water lakes in the West from northern New Mexico to southern Manitoba. The two grebes were considered the same species until 1985. They look quite similar (the black cap extends below the eye in the western, but not in the Clark's), but vocalizations (and DNA) differ between the species, and they rarely hybridize. Both types perform elaborate displays during the spring breeding season. Pairs of birds run through a series of postures, dips, and dives. The dancing culminates in the rushing ceremony resembling a bird ballet. The birds scuttle across the surface of the water, neck and body extended skyward. Seemingly defying gravity, this walking on water extends for more than fifty feet. Pair bonding continues with the weed ceremony. Males and females clasp a piece of vegetation with their beaks and bob and weave their heads in unison, eventually spiraling around one another in an avian do-si-do.

Nesting usually occurs on a floating mat or a mound of vegetation. Both sexes help raise the young. It's not uncommon for hundreds of grebes to nest on the same large lake, and after the breeding season, the birds will also congregate at staging areas during a molting period that leaves them flightless for a few weeks. They then move to wintering areas, especially along the Pacific coast and Southwest states and Mexico.

BLUEBONNETS

The state flower of Texas is actually five different species, and collectively these bluebonnets bloom across Texas in April. These lupine flowers thrive in full-sun conditions, even roadside ditches. Bluebonnets germinate

APRIL 17, 1913
Pennsylvania enacts the first hunting license requirements. The resident hunter's license fee was $1.

in the fall, overwintering as small circular clusters of leaves. Cold-hardy rosettes sprout upward as the weather warms, and they bloom between March and May. Lupines, including the Texas bluebonnets, are nitrogen-fixing plants, which help enrich soils. The heavenly scented blooms are also important host plants for a range of pollinators, including bees and butterflies. Blue is the dominant variety

Bluebonnet

genetically, but pale blue, white, and pink flowers also occur in the wild, and horticulturists have developed a number of unique types. Many people prefer to see vast fields of just bluebonnets, while others enjoy seeing red paintbrush plants splattered and sprinkled throughout the blue canvas.

It's a fitting tribute to Lady Bird Johnson that many Texas roadsides shine with bluebonnets. The former first lady was a champion for many conservation causes and pushed for highway beautification programs. In 1982 she helped open the National Wildflower Research Center near Austin, Texas, which was renamed the Lady Bird Johnson Wildflower Center in her honor fifteen years later.

> ## APRIL 18 <

WHAT'S GOOD FOR YOU IS GOOD FOR ME, OR IS IT?

Ecologists have identified five main ways in which species can interact with each other: competition, predation, parasitism, commensalism, and mutualism. For example, species that fill similar niches can directly compete with each other for habitat resources (food, water, shelter, or space). There are only so many nest cavities in the landscape, and many species are looking

APRIL 18, 1924

Chiricahua National Monument is established in southern Arizona. The area of rugged pillars is known as the Land of the Standing-Up Rocks by the Apache.

to utilize them. Predation refers to a direct impact where one organism, the predator, kills and consumes the other organism, the prey. Parasitism indicates that one organism benefits and the other is harmed but not killed. If you spend enough time outside, you (or your pets) will likely be parasitized at some point by external parasites, such as mosquitoes, ticks, or fleas, or perhaps by internal ones like tapeworms. Commensalism is when one organism benefits while the other is not impacted. Barnacles growing on marine mammals demonstrate this concept.

Mutualism is when both organisms benefit. Lichens demonstrate a form of mutualism since they are made up of a fungal and an algal (or sometimes cyanobacterial) partnership. Pollination is another type of mutualism: plants get pollinated and the pollinators get nutrients in the exchange.

> ## APRIL 19 <

NOCTURNAL PLANTS

Not all showy blooms bask in the light of day. Though the majority of plants shine during the daylight hours, a few species work on their night moves. A suite of plants reveals their petals after hours, relying on nocturnal pollinators, especially moths and bats. Plants that bloom after dark are notably abundant in tropical and desert environments. In the Southwest many plants bloom at night to avoid the hottest temperatures. Organ pipe cactus (see May 13) and other species depend on bats for pollination. In other parts of the world these nocturnal mammals are key to a number of some people's favorite foods: mango, banana, cocoa, durian, guava, and agave. Instead of elaborate and showy petals, nocturnal bloomers are often pale and rely on sweet aromas to entice pollinators, but they can still be beautiful. Their pollination is increasingly impacted by light pollution.

Many native primrose species grow throughout North America from dry prairies and dunes to wet meadows and roadside ditches. As the name implies, most species of the evening primrose open their petals as day turns to dark. Sphinx moths and bees will visit the primroses, and in some species, each flower blooms for but a single night. Especially in the southern part of the range, a few of the primrose species are diurnal bloomers.

▶ APRIL 20 ◀

WILD PIG-LIKE ANIMALS

While author and Southwest enthusiast Edward Abbey suggested that "wild pig-like animal" described some of his closest friends, the term is usually a reference to the javelina, or collared peccary. Despite the snout, javelinas aren't close relatives of hogs. They stand a couple of feet tall and weigh between thirty-five and fifty-five pounds. The name collared peccary comes from a faint band of hair around the neck that contrasts with the dark body. The species is found in Arizona, New Mexico, and Texas south to Argentina.

There are a number of structural differences between the new-world javelinas and the old-world pigs, the most obvious being that the former have short, decidedly non-pig tails and one less toe on their hind feet. Javelinas have thirty-eight teeth, including a pair of straight canine teeth. Pig species have either thirty-four or forty-four teeth with curved canine tusks. Javelinas have scent glands that are important for territory marking and individual identification. The species travels in family groups, usually of ten or less, but a group can be as large as fifty animals. Javelinas are native wildlife, but wild boars are feral pigs whose population has expanded dramatically, leaving a path of destruction in their wake.

Javelina

WATERMELON SNOW

At high elevations, snowpack can linger well into summer. These snowbanks can appear dingy as sediment mixes with the melting drifts, or under the right conditions, they can also cast a pinkish hue. Termed watermelon snow, the pink snow is created by a cold-loving alga, *Chlamydomonas nivalis*. Some early researchers rightly suspected that algae caused the coloration. Others were perplexed and attributed it to mineral deposits, fungal growth, or pollen. Pink snow has been documented on snowfields and glaciers on every continent in both mountain and polar regions. The vermilion tinge is attributed to carotenoids and is thought to help protect the species from ultraviolet radiation. Watermelon snow can absorb heat and create micropuddles that melt into pink sun cups. It is fitting that this species of algae turns the snow pink, because it also has the aroma of freshly sliced watermelon.

The subtle stink is pleasing, but avoid making a pink snow cone—the algae can disrupt your digestion and is said to have a bit of a laxative quality. The base of the snow food chain, this algae is an important nutrient for snow fleas (a.k.a. springtails; see December 28), nematodes, snow worms, and protozoans.

EARTH DAY EVERY DAY

"Earth Day worked because of the spontaneous response at the grassroots level. We had neither the time nor the resources to organize the 20 million demonstrators who participated from thousands of schools and local communities. That was the remarkable thing about Earth Day. It organized itself."

—GAYLORD NELSON, US SENATOR FROM WISCONSIN

The first Earth Day was celebrated April 22, 1970. Gaylord Nelson, a well-respected politician from Wisconsin, is credited with the idea. Senator Nelson had hoped to bring awareness of environmental concerns to local, state, and national elected officials. Nelson first announced his idea of a national environmental teach-in for college students in September 1969. The grassroots response was overwhelming, and the enthusiasm for a national day for the environment reached well beyond campuses. When the first Earth Day arrived the following April, one in ten Americans participated. The inaugural effort included two thousand colleges, two thousand community groups, and ten thousand elementary and secondary schools. Earth Day was the commencement of a number of

> **APRIL 22, 1984**
>
> Ansel Adams dies on Earth Day at the age of eighty-two. His black-and-white landscape photos are some of the most recognizable images of nature.

Earth Day

environmental policy changes implemented in the 1970s, from the Clean Air and Water Acts to the Occupational Health and Safety Act.

Local participation keeps Earth Day going year after year, and the reach continues to expand. Each spring, millions of people worldwide share a collective moment of environmental awareness. But don't let Earth Day be just an annual event. Instead incorporate it into your daily life. Recognize that your decisions have consequences, and make a positive impact for the planet. Celebrate Earth Day every day.

APRIL 23

THE SHOWIEST BLACKBIRDS

Although many migrants pass by with little fanfare, that is not the case with the arrival of the orioles. Most of these showy stars spend the winters in Mexico or Central America, and their return is often met with the red-carpet treatment from enthusiastic fans putting out backyard feeders. Orioles will utilize nectar stations but are also fond of orange halves and grape jelly.

Orioles are in the Icteridae family, so they are basically glamorous blackbirds. Structurally, there are similarities among the orioles and the red-winged and yellow-headed blackbirds. The family has long, strong, pointed bills and strong, grasping feet. The two most widespread oriole species, Baltimore (in the East) and Bullock's (in the West), were once collectively called northern oriole. Another species of the East is orchard oriole. The males are a rich dark-orange color. There is an isolated population of spot-breasted orioles established in Florida, where the species was released in the 1940s. The greatest diversity of orioles is in the West and Southwest. Scott's and hooded orioles are often associated with yucca and palms, respectively. In the US, Altamira and Audubon's orioles are both limited to Texas, but their ranges extend into Central America and Mexico, respectively.

APRIL 24

STINKHORN

Mushrooms have a tendency to materialize out of nowhere, but perhaps nothing in the fungi world is as shocking as the overnight arrival of a stinkhorn to a neighborhood. As their family name Phallaceae implies, these fruiting bodies can look downright vulgar. More offensive than their resemblance to a certain bit of male anatomy is their putrid stench. Stinkhorns emit foul odors that attract flies. Sticky secretions from the mushrooms help to transfer spores to the insects, ultimately leading to dispersal among the yards of unsuspecting neighbors. It should be noted that not all members of the Phallaceae family look phallic; however, the other shapes are far easier to overlook.

There are numerous species of stinkhorns, with the highest diversity in the tropics and subtropics. One species, *Phallus rubicundus*, is increasingly common in the United States, where it is transported in wood chip mulch.

APRIL 25

SAP SNACK

All sapsuckers are woodpeckers, but not all woodpeckers are sapsuckers. There are four species of sapsucker in North America: red-breasted along the Pacific coastal region from Alaska to Mexico, red-napped and Williamson's in the West, and yellow-bellied in the boreal north and throughout the East. Sapsuckers chisel out rows of small shallow sap wells in the trunks of trees. As the sap exudes from the bark, the birds return to slurp it up with specialized tongues that dab at the juice like a brush gathering paint.

Many other forest creatures, from porcupines to bats, will also lap up tree sap. Rufous hummingbirds seem to be closely associated with red-breasted sapsuckers and rely heavily on the sap wells, especially during nesting. The migration of ruby-throated hummingbirds coincides with the sap flows created by yellow-bellied sapsuckers.

APRIL 26

GLASSY EYES

The walleye, a large freshwater fish, is named for its opaque eyes. The tapetum lucidum, a glassy, reflective layer of pigment in their eyes, gives the species the ability to see well in the low-light conditions in deep or murky waters where it thrives. These nocturnal, toothy predators specialize in eating other fish.

LAST FRIDAY IN APRIL

Celebrated at the end of April, Arbor Day recognizes the importance of trees on the landscape.

The walleye is the state fish of Minnesota, South Dakota, and Vermont, and the provincial fish of Manitoba and Saskatchewan. Relocated and stocked widely, the species is a favorite target for many recreational anglers. The majority of commercial harvest comes from the Canadian Great Lakes. Research in Minnesota waters has shown that the timing of walleye spawning has shifted in recent years, correlating with earlier ice-out dates.

On New Year's Eve, the town of Port Clinton, Ohio, hosts a walleye drop at the stroke of midnight. It is one of at least half a dozen towns that claim to be the walleye capital of the world.

ORANGE TEETH

Beavers, porcupine, muskrats, and other rodents have orange front teeth, but it isn't from a coffee habit staining their incisors. Instead these small mammals have specialized enamel on their front teeth. The back of the incisors is softer dentine, so gnawing helps keep these teeth sharp and chiseled, like the blades of scissors. The constant gnawing wears the teeth down, but rodents have rootless incisors that grow continuously throughout the animal's life. The Latin word *rodere*, meaning "to gnaw," is the origin of the word *rodent*. Rodents lack canine teeth. The gap between the incisors and molars, called a diastema, allows rodents to pucker their lips tight while nibbling with the exposed front teeth. Worldwide rodents make up 40 percent of mammalian species diversity.

BOTH ON THE MOVE AND STAYING PUT

Having been documented on every continent but Antarctica, the green darner is one of the most widespread dragonfly species on the planet. It's also one of the most recognizable. A relatively large species, the green darner, with its brilliant green head and thorax and contrasting electric-blue-patterned tail, stands out as it cruises wetland edges. This species demonstrates a remarkable migration pattern too, or rather multiple patterns. Similar to the more familiar monarch movements, green darners' migration can be multigenerational, with adults moving north, another generation moving south, and a third generation returning north.

Other green darners remain residents. Instead of migrating as adults, they enter a diapause phase (suspended state) as nymphs, only to emerge as adults the following summer. These adults then

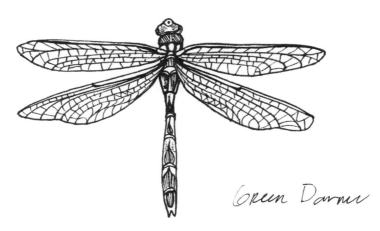

Green Darner

migrate south. Researchers used stable hydrogen isotopes locked in the wings of museum specimens and first flight data documented by citizen scientists to unlock this complex mystery of development and migration.

▶ APRIL 29 ◀

GREEN STICK

Declared the Arizona state tree in 1954, the palo verde is widespread in the landscapes of the Southwest. The name comes from the Spanish for "green stick"—the green is caused by a high amount of chlorophyll in the trunk and branches, allowing the trees to photosynthesize through the bark. Just a third of the photosynthesis happens in the leaves. These multitrunked deciduous trees grow between twenty and forty feet tall.

Foothill palo verde is the smaller of the two species found in the Copper State. Its bark has a yellowish tinge, and it is more of an upland tree. Blue palo verde, as the name implies, has bark with a blueish hue and thrives in washes. Both species produce lentil-sized seeds that can be harvested and eaten when the pods are still green and tender or toasted and ground into flour. The tree's spring flowers are also edible. Palo verde can provide microclimate benefits, including increased shade and moisture, and these conditions act as nurseries for young saguaro cactus.

▶ APRIL 30 ◀

I SEE A GOOD MOON A-RISING

The nightjar family, including whip-poor-wills, poorwills, nighthawks, pauraque, and Chuck-will's-widow, is more frequently heard than seen. These birds call out their names, sometimes incessantly, from cryptic perches. The nocturnal fliers take to the skies to sally for food, snagging insects from the air with their oversized mouths. Stiff rictal bristles were once thought to help the birds funnel flying foods down the hatch, but there is evidence that these modified feathers fulfill a sensory role.

Bright moonlight aids the nightjars in foraging, and eastern whip-poor-wills synchronize nesting to the lunar cycle. Egg-laying is timed so that hatching takes place around ten days before a

Eastern Whip-poor-will Eggs

full moon, which maximizes the moonlight and feeding time as adults are trying to feed a nest full of hatchlings. Birds of the nightjar family are sometimes referred to as goatsuckers based on old myths that they would come out at night and feed on the milk of goats.

HUNGRY PLANTS

While they may be the most famous ones, Venus flytraps aren't the only meat-eating plants. Throughout the world there are at least six hundred species of carnivorous plants. These curiosities have long fascinated naturalists, even Charles Darwin, who penned *Insectivorous Plants* in 1875. Generally these species can often be found in bogs and other habitats with limited nutrient resources. The plants still photosynthesize, but captured insects provide an essential boost, especially in the form of nitrogen. Digestive enzymes or bacteria help break down the victims.

Carnivorous plants have evolved a variety of ways to attract and collect prey. The Venus flytrap uses specialized leaves to capture unsuspecting invertebrates. Though it's not found in the Americas, the waterwheel plant is an aquatic species with similar trapping functions. A more widespread technique is found in the pitcher plants with tubular funnels, often modified leaves, that collect prey. A few species have sticky trap abilities as demonstrated by the sundews. Both aquatic and terrestrial bladderworts create partial vacuum conditions and are able to suck in passing invertebrates.

BLUE BLOODS

The curious-looking crustaceans known as horseshoe crabs are more closely related to scorpions and spiders than crabs. Similar species date back more than 400 million years, and little has changed. The four extant species of horseshoe crab are found in the Indian Ocean, along the Asian Pacific Ocean, and in the United States from the Gulf Coast and up the Atlantic to New England. The horseshoe crab's U-shaped exoskeleton protects a ten-legged body. Its long, intimidating tail, called the telson, is harmless; crabs use it to flip themselves over when they end up upside-down. They have unique blue blood because their circulatory system includes copper as opposed to iron.

Horseshoe crabs are important for commercial harvest. Their blood is sensitive to bacterial toxins, so it is used to test for contamination in medical products. After up to 30 percent of their blood is harvested, the crabs are returned to the ocean. Despite these catch-and-release techniques, horseshoe crab populations are declining range-wide (more on that tomorrow).

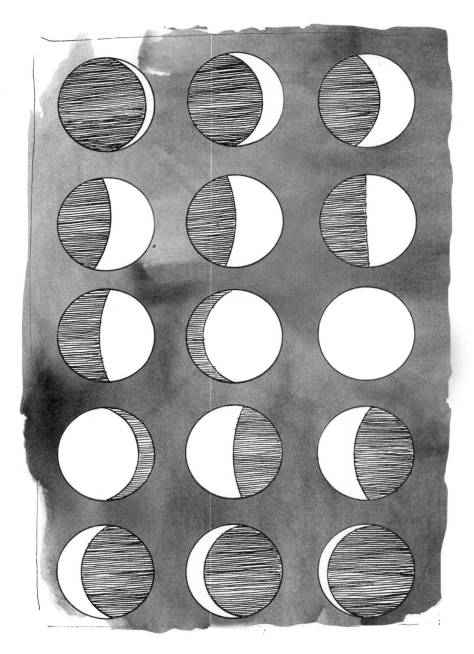

Moon Phases

GREEN EGGS AND SAND

Much of a horseshoe crab's life is spent deep at sea, but each spring, this ancient species comes to shore for breeding. Peak breeding in the Mid-Atlantic for horseshoe crabs is May, especially during high tides. The lunar cycle also impacts the timing of horseshoe crab spawning. A full moon results in higher activity. Laid in the coastal shallows by the millions, green horseshoe crab eggs are caviar for shorebirds, turtles, and fish.

MAY 3, 1906

Humboldt-Toiyabe National Forest is established in Nevada and California. It is the largest national forest outside of Alaska.

The timing of shorebird migration maximizes the value of these stopover habitats all along the East Coast. During mid-May, Delaware Bay is the hot spot for observing red knots and other migrating birds as they feast on this egg buffet during their northward migration. Red knots can double in mass at these critical stops as they journey from Tierra del Fuego in southern South America to the Arctic tundra.

NEAREST NEIGHBOR

The moon is the closest celestial body to Earth. Gravity keeps this lunar lump tethered in an elliptical orbit an average of 234,000 miles from our planet. We can only ever see one of its faces since it takes the moon 27.3 days to both make one rotation on its axis and to orbit a single revolution around Earth. This synchronous rotation keeps the same side of the moon facing our planet. There is not really a dark side of the moon, since the far side also receives light during the revolution.

The moon is illuminated by the sun and because it orbits around Earth, it shows us phases—waxing from crescent to quarter to gibbous to full, then waning to gibbous, quarter, and crescent. The new moon, when we see no moon at all, occurs when the moon is between Earth and the sun. A lunar or solar eclipse happens when the three line up just right (see September 1).

Walrus

TOOTH-WALKER

Spring is calving season in the Far North. Any day now, walrus cows will be giving birth to their roughly 140-pound babies. These marine mammals bred some fifteen months earlier on winter pack ice. Delayed implantation and a long gestation mean females produce young only every other year. In the Pacific, calving occurs on northward migrations following the seasonal retreat of sea ice. Changes in ice conditions may have profound implications on these giants, which utilize pack ice for breeding, birthing, and resting.

Walruses feed in shallows, slurping up clams and other invertebrates from the sea floor. Sensitive vibrissae whiskers provide tactile advantages, and the tongue functions as a piston to help suction up food. Tusks aren't used in feeding. Instead these gregarious animals employ their elongated canine teeth in territorial and social dominance displays. The walrus genus, *Odobenus*, means "tooth-walker."

A *NICE* TOUCH

In the 1930s, Margaret Morse Nice's revolutionary song sparrow research set a new standard for ornithology data collection. At the time, researchers focused solely on specimen collection to determine taxonomy and distribution, but little was known about the behavior and ecology of birds. Nice's approach was different. She banded the song sparrows in her Ohio backyard with a series of colored bands and studied their every move and song over the course of eight years (see October 17). Observing and documenting the wild birds allowed her to uncover their territoriality, nesting habits, learning skills, mating patterns, and ability to survive.

Her work was published in two volumes called *Studies in the Life History of the Song Sparrow* in 1937 and 1943, respectively. For the first volume, Nice was awarded the highest honor from the American Ornithologists' Union (now the American Ornithological Society), the Brewster Medal. During her career as a researcher and scientist, Nice published more than 250 scientific articles, thousands of book reviews, and seven books.

MAY 7

FROGWATCH

 A program of the Association of Zoos and Aquariums, FrogWatch USA gets citizens involved in their local communities by having them monitor area wetlands for frog and toad calls. The program takes place from February to August, depending on your location. Volunteers are required to make multiple visits with standardized sampling protocols throughout the breeding season. The unique calls of frogs and toads make them easy to identify and quantify. Amphibians are considered indicators of environmental health, so researchers are especially interested in monitoring these species. Various organizations have spearheaded FrogWatch since it was initiated in 1998. There are now more than 150 FrogWatch chapters scattered throughout forty-one states and Washington, DC. But the groups are always looking to recruit interested volunteers.

MAY 8

FROG FUNGUS

As with most of nature, habitat loss has taken a toll on frog, toad, and salamander populations. Amphibians are also especially vulnerable to slight changes in the environment. But there is another threat that is taking out multiple species: chytrid fungus. The fungus causes the disease chytridiomycosis in frogs. Across

> **MAY 8, 1926**
> Sir David Attenborough is born in London. Attenborough is the iconic voice of numerous nature documentaries.

the planet this lethal form of chytrid has been found in more than five hundred amphibian species. It is thought to have wiped out nearly one hundred species entirely. While no frogs or salamanders have gone extinct in the US or Canada due to the infection, numerous species have suffered significant declines. Chytrid doesn't affect all species equally. The American bullfrogs seem to have resistance to chytrid, although they serve as a carrier for the disease. A related ailment is impacting salamanders, although it has yet to be documented in North America.

MAY 9

COASTAL NURSERY ESTUARY

Coastal bay waters, where fresh water and salt water mix, are a main component of estuaries. These shallow waters are often sheltered behind barrier islands with convoluted shorelines. Estuaries and associated habitats range from open waters, marshes, and swamps to mud flats, sandy beaches, rocky shores, and mangrove forests. River deltas, tidal pools, and seagrass beds are found within estuaries. Thousands of species of fish, mammals, birds, and other wildlife depend upon these critical resources. Highly productive areas, estuaries are often called nature's nurseries; 75 percent of commercially harvested fish and shellfish start out in estuaries. Hotbeds for tourism and recreation, estuaries are under increasing pressures. Coastal communities are growing three times faster than inland counties. Since these bays are also at the end of the watershed, land management practices thousands of miles away affect their health.

> MAY 10 <

WMBD

The second Saturday of May is a party for the birds in the United States and Canada. World Migratory Bird Day (WMBD) has been celebrated since 2000. (Fall is when WMBD festivals traditionally take place in Mexico, Central and South America, and the Caribbean.) Environment for the Americas, the organization that oversees the program, believes in conserving birds by connecting people. Each year they choose a theme to highlight a conservation issue facing migratory birds. Recent topics have included stopover habitat, plastic pollution, and collaborative partnerships.

More than six hundred locations participate, and each celebrates in its own unique way. Many locations hold WMBD festivals. Others focus on hosting bird walks. All WMBD events are great introductions to the hobby of birding and the importance of conservation. If you're already a bird enthusiast, consider volunteering at this year's WMBD. Sharing your passion with others is one of the easiest ways to give back to nature.

> MAY 11 <

SOIL CRUST

Parts of the Southwest are covered in a thin delicate crust made up of cyanobacteria, algae, and fungi. The bacteria create a gelatinous material that cements soil particles together to form this surface. These cryptobiotic soils (*crypto*, meaning "hidden," and *biotic*, or "living") reduce wind erosion, absorb water, and reduce runoff. Associated mosses and lichen often stabilize these complex systems and provide nitrogen-fixing functions. Well-established cryptobiotic crusts take on darker appearances than the surrounding soils. Rarely more than a couple of inches thick, the layer is always quite fragile. Mature crusts can be thousands of years old, but a single pass by desert wildlife or trailblazers can disrupt the crust for decades. Cryptobiotic crust is visible in the Colorado Plateau, the Great Basin, and the Sonoran Desert—including Glen Canyon National Recreation Area, Mojave National Preserve, and Canyonlands and Great Basin National Parks.

STEP THIS WAY

When people and a number of other mammals walk, each step follows this flat-footed pattern: Heel, toe. Heel, toe. This plantigrade locomotion provides stability as an advantage. The trade-off, however, is that it is a relatively slow way to move. Bears, raccoons, wolverines, and skunks all share this full-footed meander. Digitigrades, such as dogs and cats, walk on their digits or toes. When they move, just the front of the metatarsals and the phalanges touch the ground. What we think of as our ankle is elongated and appears farther up the leg on these animals.

In unguligrade (meaning "walking on hooves") movement, just the tips of the phalanges come in contact with the ground. Even-toed ungulates include the deer family, javelina, bison, bighorn sheep, and pronghorn. Odd-toed ungulates are represented by the horses, tapirs, and rhinoceroses. In most instances, these long-legged critters are built for speed.

Grizzly Track

ORGAN PIPES

The southern border of Arizona is home to a patch of organ pipe cactus, rare in the United States. Sensitive to frost, the species is more abundant at lower elevations and lower latitudes. It grows mainly in Sonora and Baja California, Mexico. The numerous organ pipe stems grow in clumps from a single plant. The stems can grow up to about 2.5 inches per year under the best conditions. Individuals can live 150 years or more. After about age thirty-five, the cactus will finally bloom for the first time. Flowers open at night and are pollinated by bats, especially long-nosed bats.

In addition to its namesake species, Organ Pipe Cactus National Monument is home to thirty other species of cactus. The national monument and its sister park, El Pinacate y Gran Desierto de Altar, in Mexico, are both international biosphere reserves.

FLEX THOSE MUSSELS

Often overlooked pieces of nature's puzzle, freshwater mussels have some of the best names in nature: plain pocketbook, white heel splitter, fawnsfoot, Wabash pigtoe, and giant floater, for example. World mussel diversity is greatest in the Midwest and eastern regions of the United States, but populations in those places are in serious decline. Researchers suggest that nearly 75 percent of the mussel species in the country are endangered, threatened, or species of special concern. As many as thirty-five species are presumed extinct. The Mississippi River system is home to forty-nine species of mussels and four types of freshwater clams.

> **MAY 14, 1930**
> Carlsbad Cavern National Park is established. The area in southern New Mexico features more than one hundred caves.

The life cycles of most mussels require a host. The young larvae (glochidia) become lodged in the gills, fins, or bodies of a fish host. This mussel-to-host relationship can be very specific. The shovel-nosed sturgeon, for example, is the only known host of the hickorynut mussel. The fish is also host to pimplebacks

Freshwater Mussels

(another great name, right?) and yellow sandshells. Having fish hosts helps to disperse the otherwise stationary mussels. After a few days or weeks, the glochidia drop off the fish and settle to the bottom substrate of the water body. Mussels can take many years to mature and may not reproduce for up to a decade. Adults may survive sixty-five years if conditions remain stable. Not all mussels are welcome, though. The nonnative invasive zebra and quagga mussels continue to expand their ranges in North America, wreaking havoc on native ecosystems.

MAY 15

BBQ PODS

The thorny mesquite plants, a group of about forty different species, are native to the United States, Mexico, South America, northern Africa, and eastern Asia. Core ranges in the US include Texas, Oklahoma, and Arizona. Honey mesquite can be found from Louisiana to California, while velvet mesquite is mainly in southern Arizona, California, and Mexico. These leguminous plants fix nitrogen in the landscape, but in the US the mesquites have a bit of a weedy reputation. Stands have increased in density in the last couple of centuries, likely due to fire suppression. Livestock grazing has also led to an expansion of mesquite as seeds are dispersed in dung heaps.

Mesquite

The pods can be up to a foot long and each may contain up to three dozen seeds. Pods don't rupture at maturity and can remain viable for ten years if the seeds aren't consumed or destroyed by insects, fungi, or wildlife. People get in on eating, mesquite too. Seeds and pods can be roasted and ground into a flour; more commonly, the wood is burned to create the distinctive mesquite-smoked barbecue flavor.

MAY 16

BIG BIRD BEAKS

The grosbeaks, literally "big beaked," include species from two separate families of birds. Grosbeaks in the Cardinalidae family include rose-breasted, black-headed, and blue, while the Fringillidae types are the evening and pine. The Cardinalidae birds return to their breeding ranges this time of year. During migration they can be abundant at feeders, but they are more difficult to spot as summer rolls around. Rose-breasted and black-headed will hybridize in the Great Plains, where their ranges overlap. Blue grosbeaks have historically been birds of the South, but they continue to expand northward, likely as a result of forest clearing creating more shrubby and open habitats.

Winter provides a better shot at hosting evening and pine grosbeaks at feeders. In the summer they retreat to northern forests and western mountains. Their exaggerated beaks are great for cracking open seeds and nuts, but all species also feed on fruits and berries, buds, and invertebrates.

▶ MAY 17 ◀

BFF

By the 1960s, populations of black-footed ferrets were already struggling. The species was included in the Endangered Species Preservation Act in 1966 (predating the Endangered Species Act). The decline in prairie dog numbers, coupled with disease susceptibility, nearly led to the demise of the ferrets. The two-foot-long weasel with a black face mask and matching dark paws depends almost entirely on prairie dogs as a food source. Ferrets were declared extirpated in the wild by 1974, and the last of the captive stock died off by 1979. In 1981 the species was rediscovered at the Pitchfork Ranch near Meeteetse, Wyoming, by the Hogg family dog, Shep. This population of ferrets was hit hard by disease, so the final eighteen were placed in a captive breeding program that continues today.

While still endangered, there are now an estimated 1,500 black-footed ferrets living in targeted recovery areas, including populations in Wyoming, South Dakota, Montana, Arizona, along the Utah and Colorado border, and in Saskatchewan. Ferret populations are supplemented with the captive-bred animals, but wild-born kits are also increasing the population of the species.

▶ MAY 18 ◀

AGAINST THE CURRENT

Fish are often thought of as either freshwater or saltwater species, but some move freely between these habitats. Catadromous fish spawn in salt water but spend much of their adult lives in fresh water. American eels are the only catadromous species found in North America.

Anadromous fish are much more common. These species are born in fresh water, then spend much of their lives in the sea, and finally move upstream, returning to their natal waters for breeding. Examples of anadromous fish in the United States and Canada include salmon, steelhead, American shad, sturgeon, striped bass, and blueback herring. Amphidromy is rare—when fish move between salt water and fresh water for purposes other than breeding.

SONGBIRD SERENADES

The passerines, or songbirds, make up about half the species of birds, including the sparrows, warblers, buntings, finches, thrushes, and many more familiar bird families. The singing of birds is quite varied, even within the songbirds. Some barely utter a phrase. Others belt out multiple notes at once, basically harmonizing with themselves, thanks in part to split double-barreled syrinxes. A wood thrush can simultaneously sing rising and falling notes. A few, like the aptly named mockingbirds and the related thrashers, can imitate other birds, along with cell phones, car alarms, and other human noises. Often short, simple, and less musical than songs, bird calls can be helpful for individuals to keep in contact with nearby birds or to alert others of dangers.

The more elaborate songs that birds sing are generally attempts to attract mates, defend territory, or reinforce pair bonding. Males do the majority of the singing, although some females also join in on the vocalizations. A pair of northern cardinals will sing back and forth to one another. For those awake to hear it, the spring dawn chorus is a defining act of nature. For those who like to sleep in, it's the most beautiful of alarm clocks.

SPURRED BLOOMS

The official flower of the Centennial State, the Colorado columbine is just one of many species of the spurred bloomer. Columbines arrived in North America roughly ten to forty thousand years ago. The genus of *Aquilegia* has been radiating across the continent since the glaciers retreated. Over time the various species of columbines have evolved into unique ecological niches. The blue columbines are all found in northern latitudes or at subalpine and alpine habitats, where they rely on bees and bumblebees for pollination. Paler blues, yellows, whites, creams, or pinkish in color, lower elevation species are especially appealing to the hawk, or sphinx, moths. These species have higher nectar levels and longer flower spurs, matching up with the longer moth tongues. The *Aquilegia* species of the Southwest, which are highest in nectar and tend to be brighter red in color, are pollinated by hummingbirds. Hybridization does occur with many of the columbine species. The only columbine species native east of the 100th meridian (see August 23) is *A. canadensis*, which is attractive to both hummingbirds and bumblebees.

> MAY 21 <

TORNADO ALLEY

More than 1,200 tornadoes stretch across the sky and touch down each year in the United States, making it a hotbed for tornado activity. The States average four times more twister action than Europe. This funnel-cloud weather phenomenon has been documented in all fifty states, although the central states from South Dakota to Texas is known as Tornado Alley. This region east of the Rocky Mountains can experience low-pressure systems that then trigger tornadoes

MAY 21, 1987

Pioneering conservationist, herpetologist, and ecologist Archie Carr dies. A few years later a national wildlife refuge along Florida's Atlantic coast set aside for the protection of sea turtle nesting habitat is named in his honor.

as cooler air from the north mixes with warmer air from the south. Notably, states to the east and south have experienced an increase in twister frequency in recent decades. Twisters can happen anywhere at any time, but most occur during the afternoon and early-evening hours from May to early June. Peak funnel-cloud season happens a bit earlier in the Southeast and is a tad delayed in the northern plains and upper Midwest.

Dust devils are different from tornadoes. These small outbursts that form on hot clear days are simply convective updrafts. Waterspouts can form in association with tornadic activity, or they can basically be aquatic dust devils.

> MAY 22 <

BLACK, BLOND, AND CINNAMON BEARS

Black bears come in fifty shades, but they are all the same species. They inhabit a variety of habitats from the southern swamps to montane forests, from the Appalachians to Alaska. The species range also extends in the Southwest into Mexico. Black bears are classic omnivores, with upward of 85 percent of their diet consisting of vegetation. Bears can put on up to 30 percent of their body weight in the fall as they prepare to enter hibernation.

The species has eyesight similar to that of people during the daylight hours, but their vision far outshines ours after dark. They have better sniffers too, and can out-sniff a bloodhound by seven times. Black bears, especially young cubs, are quite capable tree climbers. They are also impressively quick, reaching top speeds of nearly forty miles per hour for short bursts—not bad for an animal that can easily weigh two hundred pounds. Alaska's Tongass National Forest has some of the most consistently visible black bears, although you have a chance to see one at a number of national parks and forests across the US.

MAY 23

PROJECT NESTWATCH

The Cornell Lab of Ornithology is a leader in citizen science opportunities. Project NestWatch (www.nestwatch.org) is just one program the organization oversees. NestWatchers collect data on breeding timing, nesting habits, and success rates for a variety of bird species. The program is especially popular for monitoring cavity-nesting species as they will readily use a bird box, but people also keep tabs on other nests. The Cornell Lab of Ornithology provides an expansive resource list to help train NestWatchers. They even have nest box plans suitable for many of the target birds. Participants are asked to monitor nests every three to four days until the young birds have fledged (left the nest). The Cornell Lab provides data sheets, but a growing number of participants use the mobile app to record data directly from the field.

In just one example, by coupling NestWatch data with museum documentation and fieldwork, researchers have shown that over the last hundred years, many species in Northern California have begun nesting five to twelve days earlier in the seasons. Other species have shifted their ranges on the landscape.

THIRD FRIDAY IN MAY

Bike-to-Work Day, part of National Bike Month, is celebrated across the US, with most large cities and many smaller communities participating. Originated by the League of American Bicyclists, the day encourages people to try commuting to work or school by bicycle. More than 4 billion trips are made annually by bicycle in the US.

MAY 24

HELPLESSLY YOUNG

In the nest of a typical songbird, like a yellow warbler, a clutch of nearly featherless baby birds begs for their next meal. These hatchlings are termed altricial young. It takes a bit of time and parental investment until they are capable of moving about. As the season progresses, the young birds get bigger and additional feathers grow in. By the time the brood is ready to leave the nest (fledge), they are basically the size of the adults and fully feathered. These fledglings still depend on the adults but are able to at least somewhat awkwardly move or fly. Other wildlife can also have altricial young. Baby rabbits, rodents, and carnivores depend on parental care initially. But not all wildlife is altricial.

MAY 24, 1911

Colorado National Monument is established near Grand Junction, Colorado. From Rim Rock Drive to the backcountry, the monument offers something for all nature enthusiasts.

Many youngsters are precocial, meaning that they basically hit the ground running without any parental investment. Most reptiles, for example, fall into this category. Shorebirds and waterfowl are also highly precocial in many cases; the young are quite feathered and mobile shortly after hatching. They are not completely self-sufficient, however; parents may still help with feeding and thermoregulation.

yellow warbler nest

▶ MAY 25 ◀

AMERICA'S MARSUPIAL

Australia is the epicenter for marsupials like kangaroos, wallabies, and koalas. The United States has only one—the Virginia opossum. Don't get tripped up on the name. The opossum isn't related to the nonmarsupial possums of Australia. True to their marsupial traits, opossums deliver tiny young that live for the first few months in mama's pouch. Opposable thumbs on their hind feet make it easy to identify their tracks. The naked tail is slightly prehensile, but they can't dangle from a tree limb by this appendage. Adult opossums have mouths full of fifty teeth, more than any other land mammal in North America. They will flash their teeth and hiss if they think they can't get away from trouble. Opossums are capable of delivering a bite, but generally the posturing is mostly a bluff. As a last resort the opossum will play dead. It's not really a party trick the animal can cue up on command, though; it's a physiological response to danger.

Opossums have been expanding in range northward for decades, likely in part due to urbanization and aided by warming temperatures. The species doesn't hibernate, but it does hunker down for the winter. Many northern opossums show signs of frostbite on their ears and tails.

▶ MAY 26 ◀

REGIONAL DIALECTS

Some birds innately know the song of their species. Others learn the tune by listening to the songs of others. This learning by hearing occurs during what ornithologists call the critical period, and it starts in the nest. After fledging, the young birds begin to replicate the songs they hear. The lessons continue, and practice makes perfect. This localized learning has led to regional dialects for many species of songbirds.

White-crowned sparrows are an especially well-documented case of birds with accents. Their basic call sounds like a sweetly whistled, "I . . . gotta go wee wee . . . right now." Sparrows in some areas can speed up or slow down the phrasing. In some regions, the final two trills are omitted entirely. The differences can provide a fun challenge for birders exploring a new area.

▶ MAY 27 ◀
OXBOW

From shallow riffles to deep pools, the depth of water isn't uniform below the surface. Bend after bend, as rivers meander downstream, there are hints as to what is happening underwater. Faster moving water carries more sediment than slower flows do. The inside of a river bend is the deposition zone. These shallows allow sediment to settle and gradually accumulate as a point bar. Water picks up speed on the outside of the bend, eroding a near vertical cut bank or river cliff over time. Longer-term shifts on the watercourse also occur.

Oxbows are meanders that have split off from the main channel. One of the most photographed oxbows is in Grand Teton National Park where Mount Moran is reflected in a bend of the Snake River. The broad, slow-moving pool is a great place to look for one of the park's most charismatic critters, river otters, as they fish for cutthroat trout.

▶ MAY 28 ◀
SPARK BIRD

There seem to be two types of nature enthusiasts: those that grew up spending every waking moment outside and those that came to appreciate the wilds later in life. Both are equally valid. Ask people when they first became interested in natural history, and they likely have a specific story. For bird folks, this is called a spark bird. It's that one moment that got them completely hooked on birds. Some nature enthusiasts likely have spark snakes or spark orchids. Maybe there are even spark clouds floating by.

What's wild is that the moments that ignite a passion for nature are often everyday encounters, which likely played out multiple times before but were simply overlooked. Being a naturalist is basically paying attention. Nature is all around. The best thing you can do for nature is to help light the spark for someone else.

Snake River Oxbow, Grand Tetons

fireflies

SIMULTANEOUS FLASHING

Sadly, some parts of the country don't have lightning bugs, or fireflies, but they are a highlight of summer for regions that do. That first flash of summer can instantly make anyone feel like they are eight years old again. The bioluminescent flash in the insect's lower abdomen is a chemical reaction that creates nearly 100 percent light and very little heat as a by-product. There are a number of different species of these beetles. Most shine yellow or green, while a few emit a bluish tint. The flash is generally intended to attract a breeding partner, but a few predatory fireflies use this mood lighting to attract a meal, not a mate.

One species, the synchronous firefly, has a unique pattern where multiple males can flash in unison. For years the only known population was in Great Smoky Mountains National Park. More recently a second location has been identified at Congaree National Park, in South Carolina. Both locations have to limit visitation during peak firefly season (mid-May to mid-June), but everyone should see this mesmerizing sight at least once in their life.

NATURE SPIT-UP

For many bird species, regurgitation is simply known as feeding time for the youngsters. In pigeons and doves (along with flamingos and some penguins), crop milk, a secretion from the lining of the crop, is fed to hatchlings. (The crop is a pouch near the throat used to store food temporarily.) Many species regurgitate partially digested food and then share it with nestlings. In other species, or for older young, food is basically stored in the crop and then transferred to the other individual. It can be quite endearing to see this feeding of helpless nestlings. It's somewhat less charming when a bald-headed turkey vulture projectile pukes bits of rancid roadkill, which they may do when threatened. A few birds, including vultures and many seabirds, will regurgitate as a defense mechanism.

Owl Pellet

Other species are much more dignified with their vomiting. Owl pellets are undigestible bits of bones, fur, feathers, or exoskeletons. These scraps are collected and compressed in the owl's gizzard and then regurgitated as an oval-shaped pellet. Owls aren't the only pellet producers. Other raptors, including hawks and eagles, also spit up pellets, as do some gulls, kingfishers, dippers, corvids, and shorebirds.

MAY 31

THE EYES HAVE IT

"Eyes on the side, animals hide. Eyes in front, animals hunt." The world can be an eat-or-be-eaten kind of place, depending on your place in it. For prey species, one of the first lines of defense to avoid becoming dinner is keen eyesight. Eyes on the sides of the skull provide a wider field of view, making it easier to see danger approaching from the periphery. Deer have about a 10 percent blind spot directly behind them. For predatory species, including the canids (dogs, wolves, coyotes, etc.) and felids (cats, from domestic housecats to cougars), depth perception at close range is more important. Eyes on the front of the skull help carnivores pinpoint objects, thanks to binocular or stereoscopic vision. The relative location of the eyes is just part of the story.

Pupil shape also correlates with an animal's natural niche. Wide pupils, like the rectangles of bighorn sheep and mountain goats, give prey species a broad field of view, even when eating with their heads down, while vertical pupils help predators focus at close range as they go in for the kill.

JUNE 1

POINTS NORTH

It's not the brightest, but Polaris, or the North Star, is a beacon in the night sky. The entire northern sky turns around this pivot point, the north celestial pole. The North Star is made up of a trio of stars orbiting a common center mass. It is only about the fiftieth brightest object in the night sky, but it is relatively easy to locate. If you can find the Big Dipper, you can find the North Star: the two stars on the bowl section farthest away from the handle of the Big Dipper (Merak and Dubhe) point directly toward Polaris. If you use the distance between Merak and Dubhe (so the height of the bowl, or cup) as your unit of measure, the compass star is about five times this length away from the Big Dipper. The other way to identify the North Star is to locate the star at the far end of the handle in the Little Dipper constellation.

Polaris is directly overhead at the North Pole. The farther south you go, the closer to the horizon and lower in the sky the star appears. It dips below the northern horizon south of the equator.

JUNE 2

BITE ME

For some nature enthusiasts, mosquitoes can put a damper on the outdoor experience. World-wide, mosquitoes are a vector for diseases that affect both people and wildlife, but it's important

Coyote Skull

to remember that they play important roles in the ecosystem. There are more than three thousand mosquito species worldwide and nearly two hundred unique types in the United States alone. In general, mosquitoes lay eggs in standing water, so minimizing backyard puddles can help reduce local populations later in the summer. After hatching, larvae wriggle about in the water column, rising to the surface at regular intervals to breathe air. Mosquito larvae are an important food source for a number of aquatic predators, from dragonflies to trout. The mosquito life cycle continues in the water as individuals go through the pupa stage, which lasts from a couple of days to a week or more. Next, adults emerge from the water—likely the first time people really start to pay attention to them, especially the biting females.

Very few species actively target humans, and a variety of factors are at play with the ones that do. Mosquitoes seem to key in on carbon dioxide as well as lactic and uric acids and ammonia released by people. Those with type O blood seem especially delicious for the flying needles. And in a subject that clearly needs further study, drinking beer can increase the likelihood of getting mosquito bites.

> JUNE 3 <

INSANE IN THE (NICTITATING) MEMBRANE

Eyes are sensitive and fragile, but they have built-in protection in the form of lids. A drawback to eyelids, however, is that it's impossible to see anything when they are closed. One adaptation that has evolved across taxa is a nictitating membrane. This clear eyelid shows up in a few species of birds, amphibians, reptiles, fish, and even some mammals. The translucent membrane is often found in species that dive underwater, including alligator, manatee, beaver, and diving ducks. The third eyelid, as it is sometimes called, also helps prevent dry eyes and is especially useful for fast-flying birds. For woodpeckers, that membrane serves as something like a seat belt, protecting the eye and absorbing some of the impact as they pound trees and peck wood. In the Far North, snow blindness from ultraviolet radiation is a real threat, but the nictitating membranes of polar bears are thought to protect their eyes. Domestic dogs and cats have nictitating membranes, but they lack many of the muscle fibers that control the lid; therefore it is rarely visible.

American Alligator

VANILLA BARK

Ponderosa pine (and a few closely related species) thrive in savanna-like conditions throughout the West. These long-needled pines have a characteristic reddish-orange bark that helps identify the species. The signature bark of the ponderosa pine hides a secret. Next time you find one, give it a big old whiff. Stick your nose in one of the crevices of the bark and inhale. You'll be greeted with sweet wafts of vanilla or butterscotch. The chemical composition of the sap puts off a delicious scent.

This thick bark protects the trees from fires that sweep through the understory on fifteen- to thirty-year cycles. These low-intensity burns don't kill mature trees; instead they burn off understory growth and saplings. Old photos, including some from the Custer expedition, show that conditions in areas like the Black Hills of South Dakota and Wyoming used to be far more open. In many cases, fire suppression has led to a denser canopy and thicker forest conditions. Today, land managers mimic these historic conditions with prescribed burning and mechanical thinning.

NOW YOU SEA ME

The waters of the southern United States are home to five species of sea turtles. All feed on different things and nest in dispersed areas, yet populations of each species seem to be in trouble. Historically, turtles have been exploited for meat, eggs, skin, and shells. They have also experienced heavy mortality as bycatch from fisheries, although turtle excluder devices (called TEDs) help minimize this threat now. In addition, coastal habitat destruction impacts nesting sites, and warming temperatures affect hatchling sex ratios and turtle food sources.

JUNE 5

Each year World Environment Day highlights a new theme related to the importance of doing something to take care of the earth. The United Nations first declared this day in 1974.

Loggerhead sea turtles are the most abundant nesters in the US and are found predominately on Florida beaches, especially at Archie Carr National Wildlife Refuge from May to August. The timing of loggerhead nesting is shifting earlier as surface temperatures of the ocean increases. The largest of the sea turtles, leatherbacks grow to seven feet and can weigh over one thousand pounds. They dive the deepest of all the sea turtles and feed extensively on jellyfish. They nest in small numbers from North Carolina to Texas. Kemp's ridley turtles are much smaller, weighing less than one hundred pounds. Diurnal nesters, Kemp's congregate at Rancho Nuevo Beach in Tamaulipas, Mexico, although a small population lays eggs on Padre Island National Seashore in Texas. Green

Loggerhead Sea Turtle

sea turtles, named for the color of their cartilage and fat, are herbivores and graze on seagrasses and algae. Hawksbill sea turtles cruise through the Gulf of Mexico and survive mostly by scraping sponges from coral reefs in tropical waters.

JUNE 6

THE JACK PINE WARBLER

Some species are generalists, meaning they are highly adaptable and can be found in a variety of habitats. Others are specialists. The endangered Kirtland's warbler is very much the latter. Named for early Ohio naturalist, physician, and judge Jared Potter Kirtland, the species requires young stands of jack pine for nesting. Jack pines have serotinous cones, where the seeds release following fires. The birds use these thick stands, preferring sites where the trees are between about six and fifteen years old. These early successional habitats are limited, and Kirtland's warblers face additional pressures from nest parasitism by brown-headed cowbirds.

The warblers' core breeding range is Hartwick Pines State Park near Grayling, Michigan, but isolated populations also exist in Wisconsin and Ontario. The entire population seems to winter in the Bahamas in the dense undergrowth of pine forests.

JUNE 7

CRAZY JUMPING WORMS

The United States is home to at least 120 types of native earthworm, with an additional 60 or so species of nonnatives. These exotic species are having an adverse impact everywhere. The destruction is most noticeable in forests in the upper Midwest, a location that hasn't had earthworms since the last glaciation. Crazy worms or jumping worms (*Amynthas spp.*), named after their frantic wiggling and writhing behavior, are Asian worms that are prolific breeders (they can reproduce without fertilization) and voracious eaters. Adults die off in the winter, but they leave behind tiny cocoons to repopulate the following spring.

In places like Wisconsin's Whitnall Park, crazy worms are destroying the accumulated leaf layer of the forest floor, decreasing native biodiversity, and causing a proliferation of nonnative plants. As with many nonnative species, the worms and cocoons are often transported unintentionally in mulch and compost, so it is important to be careful when moving natural products to new locations.

METAMORPHIC ROCKS

Just like metamorphosis is a big change for a cater-pillar or a frog, metamorphic rocks result from big changes as well. Rocks are classified as sedimentary (see December 2), igneous (see August 11), or met-amorphic. Metamorphic rocks are formed when a rock is changed by intense heat or pressure. The heat or pressure causes the rock to recrystallize, a process that often occurs deep within the earth. The resulting structure, along with the dominant mineral type, helps categorize metamorphic rocks. Metamorphic rocks are generally dense and strong, and they often show layers of visible crystals.

JUNE 8, 1906

The Antiquities Act, signed by Theodore Roosevelt on this date, gives presidents the authority to protect cultural, natural, and scientific features and is the basis for the creation of national monuments.

The most common types of metamorphic rocks include marble, gneiss, schist, and quartzite. Slate is a metamorphic rock that has been used as chalkboards, floor tiles, and roofs. Parts of the Appalachian Mountains, including the Green Mountains of Vermont, expose Precambrian gneiss and quartzite, rocks that coincide with the location of the ancient Grenville Mountains.

FOGBIRDS

The spotted owl has been the center of attention in the Pacific Northwest for a number of years, but another bird in these same forests has flown under the radar. The marbled murrelet is a stocky waterbird related to puffins. Like the other alcids, murrelets are coastal specialists that thrive on the open seas. The nesting habits of the marbled murrelet were unknown until the 1970s. Although most of the species in this family lay their eggs on rocky coastal cliffs and island nests, marbled murrelets nest on branches high in the treetops of coastal forests, miles inland. Adults feed in ocean waters, mostly close to shore at depths of less than one hundred feet. Both sexes feed the young birds. Loggers have called murrelets fog lark or fogbird, while fishermen once called the species the Australian bumble bee. The species is listed as threatened under the Endangered Species Act and is recognized as endangered by the states of California, Oregon, and Washington.

Marbled Murrelet

Cougar

A CAT OF MANY NAMES

Whether you call it cougar, mountain lion, puma, or catamount, *Puma concolor* is the big cat of much of the Americas. It has the largest distribution of any terrestrial mammal in the Western Hemisphere, with populations from the Yukon to the southern Andes. The species has been extirpated from much of the eastern United States and Canada, although individuals do show up from time to time. Most sightings have been of young males, and many likely originated from South Dakota. The Florida panther is an endangered subspecies still clinging to sparse habitat corridors from the Caloosa-hatchee River to the Everglades.

The cats can hit speeds of thirty-five miles per hour but most often are sit-and-wait ambush predators. Their diet is generally heavy in ungulates, such as deer and elk, but everything from mice to moose has been recorded as prey. Young cougars tend to take smaller prey. A single deer will provide a cat with around ten days of sustenance. Adults on average weigh in at around 150 pounds and are generally solitary. They are spotted when they are born, and at less than a pound in weight, the kittens are initially helpless. Juveniles will remain with their mother for two years or more.

MATCH THE HATCH

Stoneflies and related salmonflies are like popcorn to fish—they can't get enough of them. These invertebrates are also bioindicators of clean waters. Of the nine different families of stoneflies in North America, most spend considerable time as aquatic larvae and are often found in relatively cool flowing waters. The nymph stage can last multiple years. They undergo a partial metamorphosis, so eventually naiad larvae emerges from the water and molts into an adult. The slender adults, upward of two to three inches long in some species, live less than a month and are singularly focused on breeding; they don't eat.

Most species have two tails, or cerci, an important detail for anglers looking to match the hatch. Timing varies based on location, but stonefly and salmonfly hatches are some of the most anticipated natural spectacles. The hatch can start in April in coastal states and can run well into July for streams of the northern Rockies.

CONSERVATION NATION

If you've ever had the pleasure of peering across the massive Grand Canyon in Arizona or walking among the towering redwoods of Muir Woods in California, you've got Teddy Roosevelt to thank. He wasn't considered just the first modern president (serving from 1901–1909), he was also the first environmental pres-

JUNE 12, 1944

Big Bend National Park is established along the Rio Grande in West Texas.

ident, as he challenged the country's notion of unlimited natural resources. After his first year in office, Congress passed the Reclamation Act on June 17, 1902, allowing the federal government to pursue water diversion and retention projects—much needed for family farms in the arid West.

His quest for conservation continued throughout his presidency. On June 8, 1906, Roosevelt signed the Act for the Preservation of American Antiquities. Also known as the Antiquities Act, or the National Monuments Act, it gives the president authority to proclaim national historic landmarks and objects as national monuments, including sites such as Devils Tower in Wyoming, Lassen Peak in California, and Natural Bridges in Utah. Today we owe 200 million plus acres of national forests and parklands to his efforts. You can catch a glimpse of our twenty-sixth president at Mount Rushmore National Memorial in South Dakota.

PSEUDO FIR

Oregon's state tree, the Douglas-fir, is a towering conifer of the western United States and Canada and a favorite perch for nesting marbled murrelets (see June 9). Not a true fir, the Douglas-fir was named in honor of David Douglas, a Scottish botanist. Curiously, the scientific name, *Pseudotsuga menziesii*, refers to a different Scottish botanist, Archibald Menzies. *Pseudotsuga* means "false hemlock," another indication of the species' confusing taxonomy.

The tree has distinctive cones with trident-shaped bracts, and a legend to go with them. The story goes that a huge wildfire was burning through the forest, and the woodland critters were all trying to avoid it. Some could outrun the flames, but not the little mouse. Just as the fire was about to overtake the small mammal, it sought shelter headfirst in the cone of the Douglas-fir. You can still see the two back feet and little tail peeking out from the cone scales.

Douglas-Fir Cone

TONGUE TOES

As butterflies flutter by, they can capture the imagination. Try putting out a feeding station to get close-up views this year. Butterflies will visit flowers, but many species will also feed on apple slices and fruit scraps. A window feeder can be the easiest way to watch them eat. Notice the proboscis. This straw-like tube allows them to slurp up nectar, sap, and juices from fruits and coils up when not in use.

Butterflies have taste receptors on their feet. If you leave out a peeled banana, you may see them stamp around on it to taste it. These sensors are especially helpful in determining which host plants to lay eggs on. The taste buds on the toes are two hundred times more sensitive than what people have on their tongues. Keep that in mind the next time you see a butterfly perched on a pile of scat, which they do to obtain salts and amino acids that plants lack.

JUNE 15

MERMAID PURSES

Beachcombers worldwide occasionally find small pouches, usually no more than a couple of inches across. These little leathery handbags are affectionately termed mermaid's purses,
although they are also called devil's purses, perhaps for the horns
sticking off the ends. These little pockets are egg cases for skates, ray, and a few species of sharks. The egg cases are laid on the seabed or attached to seaweed and are held in place by those purse straps, or horn tendrils. The embryo develops inside and then hatches. Most egg cases you find on the beach are old and empty. Occasionally they are intact, and you can see the miniature bodies inside if you hold the case up to the light.

JUNE 16

FROM ANOTHER MOTHER

You'd think a nest would be a fairly private and secure structure. Sometimes, though, both parents leave a nest unattended, and certain avian species exploit these moments of vulnerability. Brown-headed cowbirds are an example of these nest parasites. They will lay eggs in the nests of other species of birds. Over 220 species of birds have been documented as unwilling hosts to cowbird eggs. The eggs of the intruder have a relatively short incubation period, so cowbird hatchlings get a jump start on the young of the host species. Sometimes the host birds will attempt to destroy the eggs of the cowbird or will construct an entirely new nest, but the majority of the time the birds simply hatch and raise the cowbirds as their own. Female cowbirds tend to specialize, using one particular host species, and individuals can lay more than three dozen eggs annually. Although the cowbirds are a native species, their population and distribution have increased widely as humans have altered the landscape and created more open areas and forest edges.

Cowbirds aren't the only egg dumpers out there. Wood ducks and hooded mergansers can also lay eggs in multiple cavity nests throughout a single nesting season.

JUNE 17
RHODODENDRON BLOOMS

The state flower of West Virginia, the great laurel (*Rhododendron maximum*), also called wild or rosebay rhododendron, is found throughout the state and from Alabama to Nova Scotia. In eastern North America, it is just one of the many native shrubs in the *Rhododendron* genus, which also includes azaleas. These members of the heath family are known for their showy blooms. In the case of *Rhododendron maximum*, blooms are white to pinkish. Peak blooming varies between species, but relatively speaking occurs earlier at lower elevations. Not every plant blooms every year. Every year at least a few plants will bloom, but the highest percentage of blooms seems to happen every two to four years.

In the Smokies, at elevations above 3,500 feet, the Catawba rhododendron dominates. When in peak bloom, this species gives the bald ridges of the area a brilliant purple wash. In Great Smoky Mountains National Park, rhododendron hot spots include Chimney Tops and Andrews Bald. The Blue Ridge Parkway and Roan Mountain State Park also provide memorable rhododendron views. A different rhododendron, the Pacific, or coast, variety, is the official flower of Washington state.

JUNE 18
WATER-QUALITY MONITORING

There are hundreds of organizations involved with water-quality monitoring. Some groups are doing independent work, while others are linked in with national networks. The US Environmental Protection Agency, state natural resource agencies, and nonprofit organizations all spearhead efforts, and they all rely heavily on citizen science volunteers. Water monitoring can be labor intensive as sampling protocols often require regular visits to multiple bodies of water at numerous locations in an area. And while the water-quality data is important to obtain baseline information and to track changes over time, in many cases, the community outreach becomes equally as important when folks passing by stop to learn more from volunteers. These frontline environmental education opportunities can build the foundations of community support.

▶ JUNE 19 ◀

PINK FLAMINGOS

Lawn art can be found just about anywhere, yet Florida seems to be the stereotypical locale for pink flamingos in the yard. But what about feathered flamingos in the flesh? Were there ever any flamingos, for instance, in Flamingo, Florida? Historical records indicate flamingos were fairly common in southern Florida, but like many birds with decorative feather plumes, they were exploited heavily in the 1800s. Throughout the 1900s and into the 2000s, an occasional rogue flamingo showed up in the Sunshine State, and once, a flock of 147 flamingos showed up in Palm Beach County. In some instances, these random birds likely represented escaped captive birds, but birds from the Bahamas, Cuba, and Mexico are also potential sources of vagrant flamingos. Birds banded as chicks in Mexico have shown up in Florida, proving that free-range flamingos do still exist in the state, but breeding populations have been extirpated from Florida. Texas also occasionally sees a flamingo, probably a drifter from the Yucatan Peninsula.

Flamingos aren't the only large pink birds along the coasts surrounding the Gulf of Mexico and Florida's Atlantic coast. The roseate spoonbill is much more abundant in this area.

▶ JUNE 20 ◀

FLOATING MOONS

Moon jellies, a name applied to numerous species, are found worldwide, drifting along in both temperate and tropical waters. They are named for their round, translucent bodies. Four oval-shaped organs are often visible inside the translucent bell body. These aquatic invertebrates can reach fifteen inches in diameter and come in various shades. Coloration can be impacted by the diet; heavy crustacean intake can give the jelly a pink or lavender tint, although a diet heavy in brine shrimp can cast more of an orange hue.

Predators, including sea turtles and birds, eat moon jellies. Tragically, each year many animals die from ingesting plastic bags they mistake for jellies. An unbalanced ecosystem, especially waters high in nutrients and low in oxygen, can cause jelly blooms that lead to overabundances of the invertebrates.

SUMMER

Summer Solstice

SUMMER SOLSTICE

The Summer Solstice, the longest day of the year in the Northern Hemisphere, is near—it may even be today. While all days are twenty-four hours long, this one has more sunlight than any of the others for those in the Northern Hemisphere. This is because of the tilt of Earth on its axis. Earth's 23.5-degree tilt determines the seasons, and on the Summer Solstice, the sun's rays shine most intensely at the Tropic of Cancer. That makes it summer in the North but winter in the Southern Hemisphere. This tilt also affects the length of daylight throughout the year. You may not notice Summer Solstice if you live on the equator, where the sun shines around twelve hours a day all year long.

WINNOWING SNIPE

Wilson's Snipe

Growing up, you may have gone snipe hunting by taking your flashlight and gunny sack out after dark to call in a seemingly mythical beast. June is the season for tracking down the real snipe: a plump shorebird in the sandpiper family. Across the northern regions of the continent, from eastern Canada and the Great Lakes to the western Mountains, Pacific Northwest, and north to Alaska, snipe put on an aerial breeding display high above the wet meadows they call home. Wilson's snipe fly in territorial loops by circling high up in the sky before diving back down to the ground. These courtship flights are called winnowing. The birds create a hollow, haunting sound. It's not a vocalization; this fluttering crescendo of sound is caused by air whispering through outspread tail feathers.

In the wintertime, look for the mottled-brown birds probing for invertebrates in rice and sugarcane in the southern states.

PUFF STUFF

Puffballs are spherical fungi that release clouds of spores not from gills, like other mushrooms, but from an internal mass called a gleba. The term *puffball* doesn't mean much taxonomically. The characteristics are found in a variety of mushroom families. The giant puffball, *Calvatia gigantea*, is what you likely imagine when you conjure up thoughts of puffballs. Basketball-sized or larger, this white fungus dots the landscape from coast to coast and in temperate areas across the world. It grows in open fields and meadows or in deciduous forests.

A few people enjoy puffball steaks, but there are look-alikes, so be careful if you're foraging for these fungi. Young puffballs can have firm flesh that soaks up flavors, but the brown powdery spores on more mature mushrooms are less than palatable.

FLYING MAMMALS

Flying squirrels don't fly (see March 1), but another group of small mammals do: the bats. Bat diversity is impressive, with nearly 50 species in the United States alone and more than 1,300 kinds worldwide. A bat wing is a highly adapted hand with extended finger bones that are connected with a patagium, a membrane of skin. Bats fill a number of ecological niches. Some, like the long-nosed, are pollinators. Others, including pallid bats, are highly carnivorous. The majority of the most familiar bats—*Myotis* browns, hoary, and red, for example—use echolocation to feed on flying insects. The Mariana fruit bat of Guam, American Samoa, and the Northern Mariana Islands is a large species of bat known as a flying fox that eats fruit, flowers, and leaves.

If you want to spot bats, look for some species fluttering over small ponds as nightfall sets in. Many caves, mineshafts, and other bat hibernacula are off limits to people, but Carlsbad Caverns National Park, Bracken Cave near San Antonio, Texas, and the Congress Avenue Bridge in Austin, Texas, all provide excellent bat-viewing spectacles. An emerging threat that continues to spread, white-nose syndrome is a fungal disease that disrupts hibernation and decreases the survival rates of numerous bat species.

Hoary Bat

WHO'S YOUR DADDY?

The same DNA technology that helps you map out your heritage allows researchers to unravel paternity in the natural world. A common refrain is that many birds mate for life, but that seems to be less common than was once thought based on fascinating results from paternity tests. Perhaps as little as 10 percent of avian species are monogamous. Extra-pair copulation, on the other hand, occurs in a number of bird groups. Evolutionarily, having multiple fathers for a single brood maximizes the likelihood that all eggs will be fertilized and can improve overall genetics. Males can also gain genetic fitness by dispersing genes more widely in the population. In general, higher paternal investment lowers the likelihood of extra-pair copulation, but there are exceptions.

JUNE 26

CUTTHROAT WORLD

During the annual cutthroat trout spawning season, Fishing Bridge in Yellowstone National Park is a great place to watch fish. (You can't fish from the bridge anymore, but fishing is allowed farther upstream.) In addition to passing on their genes to future generations of cutthroats, these native trout support an extensive food chain as they move from Yellowstone Lake into the streams. At least twenty species, including grizzly bear, river otter, and osprey, feed on the fish as they congregate in the shallow waters.

The balance of this ecosystem has been thrown out of whack with the illegal introduction of nonnative lake trout into the Yellowstone Lake ecosystem. Lake trout are voracious predators of

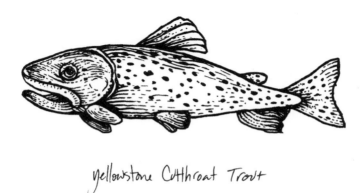

Yellowstone Cutthroat Trout

cutthroats, and the native populations have plummeted. This decline has had impacts throughout the park. Lake trout don't move into the tributary streams for breeding, and they are not a replacement for the cutts. Hybridization between nonnative rainbow trout and cutthroats is also a concern for fisheries managers in Yellowstone. The name cutthroat is for the orange markings under their mouths, but it's also an appropriate persona for a species battling so many threats to survival.

JUNE 27
LOONY BIN

Wailing calls of loons represent summer in the north woods and across Canada. Common loons are represented on the Canadian dollar coin, and they are the provincial bird of Ontario. Similarly, the species is the state bird of Minnesota and is highlighted on the quarter for the Land of 10,000 Lakes. Unlike most birds, loons have solid bones, which helps them dive deep and remain underwater for up to three minutes, although most dives are shorter. Both parents care for the chicks, and young loons can spend up to 65 percent of their time riding on the backs of adult birds during their first week out of the nest. Later in the summer, adults migrate south first, leaving the young, who follow a few weeks later.

Loons mostly winter on the coasts. They molt into a plain gray plumage, though, so don't expect to see black-and-white checkerboards swimming about during that season. There are five species of loons in North America: common, Pacific, red-throated, yellow-billed, and Arctic. In the UK, they are called divers.

JUNE 28
SWARMS OF FISHFLIES

For a few short days each summer in some regions, in scenes reminiscent of clouds of soil blocking out the sun during the Dust Bowl era, the air gets thick with mayflies. Swarms of mayflies are especially impressive in the Great Lakes region. Called fishflies or shadflies locally, these *Hexagenia* mayflies swarm by the millions. The rebound in *Hexagenia* is a sign of improved water quality in the area—even if it brings short bouts of annoyance.

This adult stage of the mayfly life cycle is short-lived: The insects mate, the females lay eggs, and then they quickly die off, leaving behind drifts of bodies. The majority of the time, mayflies live as nymphs burrowed in the sediments along the bottom of water. Nymphs typically have a two-year life cycle, although it can stretch to three or four years in colder climates.

A PROMISCUOUS POLLINATOR

Pollination is the process of plant fertilization, where pollen grains are transferred from a flower's anther to its stigma. For some species this fertilization can occur within the same plant, while for other species this pollen must move between individuals. Wind is a common dispersal technique for plant pollination, and animals also play a critical role in pollination. A number of insects are pollination specialists, as are hummingbirds and even bats.

Shooting Star

One particularly fascinating style of pollination is demonstrated by the fairly widespread plant called shooting star. This flower's petals project backward, giving the plant the appearance of a meteor blasting through the night sky. The reverse petals of this flower (and other promiscuous pollinators) allow its reproductive bits to be fully exposed. The pollen is released from the anther of these buzz pollinators by the flapping wings of passing bees. Shooting star species are found from forest understory to mountain meadows. Southern populations tend to have pinkish-white petals, and northern and western plants have darker purplish-pink hues. A number of wildlife species, including hikers, have been seen nibbling on these radish-flavored wildflowers.

THE TIMING'S JUST NOT RIGHT

Phenology (see January 1) refers to the interrelated connectedness between species. Biotic (meaning "living") and abiotic (meaning "nonliving") factors play out year after year. This ecological dance represents the rhythms of Earth, but the beat is increasingly out of step for many taxa. Flowering dates don't always coincide with insect hatches, and migratory birds, in turn, miss out on a critical food source. Similarly, leaf out is occurring later in parts of the South because plants aren't experiencing essential stretches of cooler weather in the winter months. And amphibian and reptile cycles are being disrupted in spring and fall. Some species are adapting to these changes, while others are not. Natural resource managers and backyard nature enthusiasts alike are increasingly concerned about these phenological mismatches, or asymmetry, and this growing awareness is shining a light on the role of phenology and citizen science in conservation.

JUNE 30, 1864

Yosemite Valley becomes the first area protected for preservation and public use. The action, by President Abraham Lincoln, predates the establishment of Yellowstone, the first of the national parks.

JULY 1

ONE GOOD TERN

The largest terns, Caspian, are found on every continent with the exception of Antarctica. These hefty terns fly with strong wing beats. Their bright-reddish-orange bills are thick and quite obvious, even when the birds are in flight. For much of North America, spring and fall migrations are the best times for spotting one as they have fairly restricted breeding distribution. On the border of Oregon and Washington, the largest breeding colony has taken over an artificial island in the Columbia River estuary.

Like other terns, Caspians are fish eaters that plunge beak-first into the water to snatch snacks. Initially, terns predated heavily on endangered salmon and steelhead in some locations. By placing waving streamers on potential nesting locations and using other techniques, land and fisheries managers have helped colonies of terns relocate to reduce their effect on endangered fish species.

Caspian Tern

JULY 1, 1908

President Teddy Roosevelt creates more than a dozen national forests, including Mendocino (California), Kaibab (Arizona), Deschutes (Oregon), Nez Perce (Idaho), and Ashley (Utah). During his presidency, he set aside more than 100 million acres as national forest land.

JULY 2

THE FLUTTERERS

People often assume that butterflies are brightly patterned and moths are drab and nondescript and that butterflies are diurnal and moths are nocturnal, but that isn't always the case. The easiest way to distinguish the two types of flutterers is to look at the antennae: Butterflies have straight antennae that end in little knobs. Moths have simple threadlike antennae or ones that appear almost feathery in appearance.

In the United States there are more than seven hundred species of butterflies (including two hundred kinds of skippers). There are well over ten times more species of moths with more than ten thousand types identified in North America north of Mexico. Butterfly and sphinx moth watching is popular during the day. Nighttime observing takes some special equipment though (see July 21).

BACKYARD WORLDS: PLANET 9

The National Aeronautics and Space Administration (NASA) hosts a number of citizen science opportunities, most of which take place from the comfort of your couch. In some projects, volunteers help researchers map out solar objects like galaxies, the moon, or the surface of Mars. In others, like Backyard Worlds: Planet 9 (www.backyardworlds.org), citizen scientists help discover new objects in the solar system. Computer programs can scan vast quantities of data, but the human eye is the best tool for recognizing subtle motions in astronomical images. After a short online training, volunteers sort through the Wide-field Infrared Survey Explorer (WISE) mission images from NASA to help document moving celestial bodies, including brown dwarfs and a potential planet nine.

LET THERE BE BIOLUMINESCENCE

In a number of magical places on the planet, the waters shimmer with nature's fireworks. Single-celled dinoflagellates (predominately algae) are responsible for this sparkle. Well, technically, enzymes create the flash when the organisms are disturbed. It is thought that this burst of light may startle would-be algae eaters and that the glow could draw the attention of predators, and their presence causes the algae consumers to move out of the area, thus protecting the dinoflagellates. The chemical reaction that creates the bioluminescence seems to have a circadian cycle. As night falls, the glow potential increases. Puerto Rico is especially noted for its glowing bays, but the natural phenomenon occurs throughout the oceans.

Bioluminescence isn't confined to dinoflagellates or to the seas. Fireflies also create light, as do some marine crustaceans, jellyfish, worms, fungi, a number of fish species, and a few sharks.

Bioluminescence

BANDED-TAIL BANDITS

The raccoon gets the unfortunate nickname of masked bandit. They can certainly raid a bird feeder or a garbage can, but *bandit* seems a bit harsh. Nobody calls the closely related coatis bandits. Perhaps this is because coatis don't hide behind a mask. Their diurnal habits also make them more acceptable in the eyes of many humans. In the US, coatis are limited to the Southwest, predominantly Southern Arizona. For food, coatis root around, searching out grubs, lizards, snakes, and whatever else they can find on the ground. The agile mammals are also arboreal and forage for nuts, berries, and bird eggs in the canopy. They are mostly associated with oak and sycamore canyons and move to lower elevation riparian zones in winter.

Another raccoon relative, the ringtail, is rarely seen, thanks in part to its extremely nocturnal lifestyle. Not much larger than squirrels, these impressive climbers and leapers use their semiretractable claws and ability to rotate their hind feet backward to barrel headfirst down trees and cliffs. Occasionally referred to as the ringtail cat or miner's cat, Arizona's state mammal can be found as far north as southern Oregon.

CROSSED

Hybridization is well documented in nature. The breeding of two separate species often results in the offspring being sterile. A hybrid can show characteristics of one or both of the parents. About 10 percent of bird species have been documented mating with another species. In general, more closely related species are more likely to hybridize with one another. Waterfowl seem to hybridize fairly regularly as do hummingbirds. Another group of birds that cross more frequently are the large gulls. Gulls are tricky enough for people to identify, but many different hybrid pairings have been seen.

A pair of warbler species that hybridizes frequently (leading to debates about if they are the same) are golden-winged and blue-winged warblers. Genetics work shows that they are 99.97 percent similar. The two hybridize freely and the offspring are readily recognizable, so much so that early naturalists thought these hybrids were separate species: the Brewster's warblers and Lawrence's warblers. The Brewster's warblers show the genetically dominant traits, while Lawrence's warblers exhibit the recessive traits. These hybrid individuals can breed with either golden-winged or blue-winged warblers and may also mate with another hybrid (either Brewster's or Lawrence's).

AN AMERICAN GECKO

Geckos are the predominant lizards in some parts of the world—less so in North America. The western banded gecko is the most widespread species in the United States, but even it is found only in pockets of the Southwest, including in California, Nevada, Arizona, New Mexico, and Utah. Generally nocturnal, they are often spotted when crossing roads at night or loitering near porch lights, looking to catch a meal. Geckos have specialized toe pads that, coupled with van der Waals forces, add to their clinging ability. They can easily detach their plump tail; however, when a replacement tail grows back, it is shorter and they are unable to dislodge it. Female geckos lay soft-shelled eggs that firm up when exposed to air. Western bandeds have eyelids, but many gecko species lack them; instead they are forced to lick their eyeballs clean.

Making chirps or little barks, geckos are some of the more vocal reptiles. Banded geckos often let out a little squeak when handled.

CROOKED WOOD

In mountains worldwide, the subalpine zone, just below tree line, is a brutal band of existence. Slopes are windswept, and plants here take a near constant battering. Many species take on dwarf shapes, hugging the ground as the winds whip by. Trees grow gnarled under these conditions, forming krummholz stands with low boughs sprawling near turf level. The term comes from the German word *krumm*, meaning "crooked, twisted, or bent," and *Holz*, meaning "wood." Coastal forests can also take a beating from continual prevailing winds. Here, too, trees can show characteristic flagging. The branches on the windward side are damaged, creating an asymmetry as branches grow only on the protected side of the trunk. In Newfoundland this is referred to as *tuckamore*. Look for these twisted trees above Going-to-the-Sun Road in Glacier National Park, approaching the summit of Colorado fourteeners, and on Mount Katahdin in Maine, among other windswept locales.

JULY 1906

The Mountaineers is established to facilitate exploring the wild areas and peaks of the Pacific Northwest and beyond. Today the nonprofit organization creates opportunities for people of all ages to get outside, learn new skills, conserve land, and connect with a community that loves the outdoors.

INTERTIDAL ZONE

The intertidal zone is the interface of ocean and land, the area exposed at low tide and covered at high tide. This thin swath of diversity, which rings the continents, ranges from just a couple of feet wide in places to over a mile wide in the Bay of Fundy, home of the world's greatest tides. Intertidal marine life must adapt to continually changing conditions from wet to dry. Low tide zone specialists are mostly suited for aquatic life, although they experience dry conditions for short durations.

Many seaweeds, urchins, sea stars, and anemones live in the low to middle tide zones, where tide pools provide a residual supply of water and shelter from crashing waves. High tide zone specialists are hit by pounding waves and open-air conditions. On rocky coasts, barnacles and limpets can close up and retain the dampness they need to survive. On sandy coasts, critters burrow beneath the surface to survive the dry periods. Above the high tide zone is the splash zone, where few species live.

AN EAST-WEST DIVIDE

Sometimes birds formerly lumped into a single type are reclassified into separate species. For example, what was once called northern oriole is now known by two different names, Baltimore and Bullock's orioles. Similarly, rufous-sided towhee is now recognized as eastern towhee and spotted towhee. This taxonomic split can be based on differences in plumage, vocalization, or genetics, and the Rocky Mountains serve as a natural divide between many of the eastern and western species. The distinction is more apparent in species like indigo bunting, an eastern species, and Lazuli buntings of the West, whereas species like eastern and western meadowlarks or ruby-throated and broad-tailed hummingbirds look quite similar. For some of these types, hybridization (see July 6) is common along the Great Plains, where the two species overlap. In other instances, there is virtually no hybridization. The birds can tell each other apart even if we struggle to pinpoint differences. Not all species are separated out along this east-west divide. Even though the northern flicker is the same species across the country, the red-shafted type in the West and yellow-shafted in the East look different.

PERMAFROST

A subterranean band near the poles remains virtually unchanged from season to season. While surface temperatures fluctuate over the seasons, this permafrost layer of soil, ice, and other matter that

Intertidal Zone

is found between a few inches and nearly five thousand feet below the surface remains unthawed for multiple years consecutively. Nearly 25 percent of the landmass in the Northern Hemisphere is permafrost ground, but, exacerbated by a changing climate, an increasing amount of permafrost is thawing. Alaska and other Arctic regions have recorded warming at twice the rates of the Lower 48 states in the last century: Summer temperatures have increased 3.4 degrees Fahrenheit on average, and winter averages are up more than 6 degrees Fahrenheit. Longer and drier summers result in increased frequencies of drought and fire. And as the active layer penetrates deeper into the permafrost, decomposition by soil microbes leads to the release of greater amounts of carbon dioxide and methane, greenhouse gases that contribute to climate change. The thawing buckles the land and can create thermokarst lakes in collapsed divots. Ongoing thawing leads these puddles to drain.

JULY 12
ERUPTING SPRINGS

Yellowstone National Park has exceptional geothermal hot springs, many of which erupt in the form of geysers. Home to 60 percent of known geysers, the park has the highest concentration of geysers in the world. Steamboat is the tallest, shooting water more than 300 feet into the air, but its release is unpredictable. Old Faithful is far more punctual. This icon shoots off every 35 to 120 minutes. The duration of the most recent eruption is used to estimate the timing of the next release. After longer eruptions, the geyser takes longer to recharge. At its peak release, Old Faithful spouts between 90 and 184 feet in height. Cold-water geysers, whose pressure stems from carbon dioxide, are even more rare than hot-water ones. Soda Springs Geyser in Idaho and Crystal Geyser in Utah are two examples. The eruption of the Idaho geyser is regulated by a time-release valve controlled by the city named for it and set to go off every hour.

JULY 13
STUD PUFFIN

In North America, puffin species include the Atlantic on the East Coast and both horned and tufted puffins in the Pacific. The Atlantic puffin is the official bird for Newfoundland and Labrador, Canada, where Witless Bay is home to half of North America's breeding population for the species. In addition to puffins, the alcid family is full of portly duck-like birds, such as guillemots, murrelets, and auklets (taxonomically, the rhinoceros auklet is technically a puffin). Alcids are capable of flight, but flying consumes a lot of energy because of the birds' structure. Their wings are designed to help propel them underwater, where they are efficient fish predators. Historically, the alcids were targeted by egg

Atlantic Puffin

collectors and market hunting. The flightless great auk was extinct by the mid-1800s (see November 5), and the Atlantic puffin was nearly extirpated from the United States too.

Beginning in 1973, young puffins were relocated to Eastern Egg Rock Island off the coast of Maine in an effort to reestablish a nesting colony in this historic habitat. Wooden puffin decoys were also used to recruit birds to this nesting habitat. By 1981 the first puffin nesting in decades was documented on Egg Rock, and now more than 150 pairs utilize the island.

JULY 14
SPIDER WASPS

As their name implies, tarantula hawks (wasps not hawks) are impressive predators of spiders. Like other parasitoid wasps, they sting their prey species of choice, which paralyzes the victims so that the hefty two-inch-plus wasps can haul them off to a brood chamber burrow, cavity, or mud cell. Females lay single eggs in the living prey, which eventually serves as food for the wasp larvae. When not piercing arachnids, these wasps are generally nectar eaters, and a few species will consume the so-called honeydew secretions of aphids. Their sting is reportedly quite painful (hopefully you never have to find that out firsthand), but unless you're a spider, tarantula hawks are generally docile and not likely to sting people when left alone.

There are fourteen species of *Pepsis* tarantula hawks in the United States. The most widespread, *Pepsis thisbe*, ranges from Nebraska to Texas. The species' wings have a bright, burnt-orange color, and the body can look bluish black. The females' antennae are more curved than those of the male.

JULY 15
LEAVES OF THREE

Poison ivy has a remarkable variety in growth. It can be found in open fields or in dense shaded thickets. The plant can be sprawling like a ground cover or like a vine clinging up tree trunks, fences, or buildings. The trio of leaflets that make up each leaf can be smooth edged or toothed. The leaves are often somewhat glossy and look reddish in spring and can be bright red, orange, or yellow in the fall. Poison ivy

MID-JULY

Latino Conservation Week, a program of the Hispanic Access Foundation, is held annually to support the Latino community getting outdoors and helping protect natural resources.

berries are white and are a favorite food for many species of birds. But the plant is less popular with most people. Between 70 and 90 percent of folks are allergic to the plant, or, more precisely, to the urushiol oil within it. (This is the same compound found in poison oak and poison sumac.) The mantra "Leaves of three, let it be" is a fairly helpful mnemonic, although not 100 percent foolproof. Plenty of other species also show a trifecta of leaves (or leaflets, as is technically the case for the potent plant). In winter, when leaves are largely absent, the phrase "Hairy vine, no friend of mine" is the chant of choice. The plant's fuzzy appearance this time of year is from the tendrils that help anchor the vine in place.

JULY 16
K. RATS

Summer nights are prime time for viewing kangaroo rats. Look for these plump-bodied rodents bounding across roads in the western and southwestern states. The most widespread is the Ord's, with a range extending to southern Saskatchewan and central Oklahoma. Kangaroo rats can reach lengths of a foot or more, but two-thirds of that is its tail. Hopping along on oversized hind feet, they can cover nearly ten feet in a single bound to avoid predation by owls, snakes, badgers, foxes, or coyotes. These nocturnal rats avoid the midday heat by staying in their underground burrows.

They do not need to drink water; instead they get it from the seeds they eat. Their highly efficient kidneys minimize water loss to urine, and they pee out highly concentrated crystal-like urine.

JULY 17
WRACK LINE

As the tide recedes, a bit of the sea is left behind in the form of a wrack line, which can include anything from kelps, seagrasses, and invertebrates to animal carcasses. This pile of ocean matter is an important food source for a number of critters. Coastal shorebirds, like sanderlings, often pluck snacks from the pile, including flies, beach fleas, and other invertebrates. The tangle of matter also provides cover for crabs or other terrestrial beach life. Wrack lines, sometimes called drift lines, can act as a base layer for coastal dunes by trapping wind-blown sand. Decomposing matter adds nutrients to the soil and can jump-start plant growth.

It's always neat to find driftwood, but storms can churn up the ocean and increase the volume and diversity of material in the wrack-like piles. Experienced beachcombers know to look for the most unique treasures on walks after storm surges, although unfortunately, human litter is increasingly a component of wrack lines. Doing your part to clean up beach trash will go a long way toward maintaining a clean and safe environment for wildlife in the water and on the shore.

THE SMITHSONIAN'S FIRST WOMAN CURATOR

Noted herpetologist Doris Mable Cochran started her career at the War Department and moved to the Smithsonian's National Museum of Natural History as an aide in the Division of Amphibians and Reptiles in 1918. While working, she earned a PhD at the University of Maryland, where her research examined blue crab musculature. She rose through the ranks at the Smithsonian and eventually became the museum's first female curator in 1956. Cochran was the second person elected as a distinguished fellow of the American Society of Ichthyologists and Herpetologists.

Her work focused primarily on the West Indies and Central and South America. She wrote ninety research papers describing 125 species and subspecies, especially frogs and snakes, as well as two books, *The Herpetology of Hispaniola* in 1941 and *Living Amphibians of the World* in 1961. A talented artist, Cochran illustrated much of her own work as well as that of many of her contemporaries.

THE COFFEE WEED

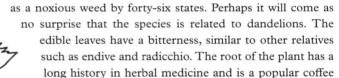

Native to Eurasia, chicory has become widespread and naturalized across much of North America. Despite the plant's showy nature, it is best to avoid planting chicory. Less invasive than some plants, the species still continues to spread. The light-blue to purple flowers are widely seen in roadside ditches, and the species is classified as a noxious weed by forty-six states. Perhaps it will come as no surprise that the species is related to dandelions. The edible leaves have a bitterness, similar to other relatives such as endive and radicchio. The root of the plant has a long history in herbal medicine and is a popular coffee and beer additive. The stem bleeds a milky latex when broken.

Chicory

UNDER THE BLACK LIGHTS

There are about two thousand species of scorpion in the world, but only about twenty-five have venom potent enough to harm a human adult. The bark scorpion is one of them. Bark scorpion ranges from Southern California, Arizona, and New Mexico south into Baja California, Sonora, and Chihuahua, Mexico. Young scorpions hitch rides on the backs of their mothers for the first couple of weeks of life. Predatory by nature, these arachnids target mostly invertebrate prey, including insects, spiders, centipedes, and other scorpions. Scorpions use pedipalp pincers to grasp prey and then dispatch dinner with a zap from their stinger. Despite having a venomous stinger, scorpions are eaten by a variety of nocturnal predators like tarantulas, lizards, owls, bats, shrews, and grasshopper mice.

One trick for finding scorpions is to use black lights to expose them. The hyaline layer of its exoskeleton shines when exposed to ultraviolet light, and this sensitivity is thought to help the scorpions seek out the darkest nooks and crannies. Fossilized hyaline can fluoresce, but freshly molted scorpions don't glow. It's unclear to scientists whether hyaline is a byproduct of the hardening process or if it is secreted after the scorpions molt.

MOTHING

National Moth Week (www.nationalmothweek.org) is celebrated the last week of July, a prime time for insect diversity across much of the Northern Hemisphere. Moths aren't just little cryptic brown fliers (although those are important species too). They are also spectacularly marked and, in some cases, quite brightly colored. The giant silk moths, such as lunas and cecropias, can be six inches across. Underwing moths, on the other hand, are a couple of inches long and have hind wings that flash pinks and oranges, making them look like vibrant watercolor paintings.

Black lights and mercury vapor lights are especially effective at attracting moths and other nocturnal insects, although just leaving your backyard light on can draw moths in. Stringing up a white sheet and shining a light on it is an easy way to get close-up views. Another technique is to paint a bait of mashed bananas and beer on tree trunks (which may discolor the bark, but won't hurt the tree).

ARMADILLO HIGH JUMPERS

If armadillos competed in the Olympics, they'd participate in the high jump. Instead of using the Fosbury Flop technique, these armored mammals jump straight vertically when startled. It is thought that this is a predator-avoidance trick, but unfortunately, the move makes them especially vulnerable to becoming roadkill. The nine-banded armadillo wasn't documented north of Mexico until 1849, when one was found in Texas. The species continues to expand its range to the north and east and can now be found well into the Midwest and Mid-Atlantic regions. Cold winter temperatures may prove to be a limiting factor into the expansion.

Predominately insectivorous, nine-banded armadillos have a long, sticky tongue for slurping up snacks. Modified claws make them efficient diggers and they can burrow tunnels more than twenty feet long. The species found in North America can't curl up into a ball, but their strong keratinized scales provide ample protection from most predators. Curiously, armadillos are susceptible to leprosy. They maintain a body temperature of 90 degrees Fahrenheit, which provides suitable conditions for hosting the disease.

VACCINIUM SPP.

There are several species of *Vaccinium* bushes, and they produce some of the tastiest fruits around. Cranberries, blueberries, huckleberries, whortleberries, and lingonberries all belong to this group in the heath family. Watching a bear use its agile tongue to strip berries from the plant is a nature event that everyone should experience firsthand at least once (and from a respectful distance). If there are no bears in the berry patch, you may want to gorge on these treats.

Found in mostly northern and mountainous zones, these bushes' growing seasons and conditions vary across the continent. Some species prefer bog habitats, while others grow as forest understory. Most prefer acidic soils. Peak berry picking is a prime time to be alive, and gathering these delicious fruits is perhaps one of the easiest ways to forage. The species are fairly easy to identify, but be sure about your identification skills before you eat them. In addition to growing in the wild, many of the *Vaccinium* varieties are widely cultivated.

Nine-Banded Armadillo

UNLIKELY GULL

In the year 1848, Utah crops were being destroyed by an infestation of Mormon crickets, a flightless species of katydid. An unlikely hero in the form of flocks of gulls swooped in to save the day. (This wasn't a flock of seagulls, mind you. Despite the fact that the word *seagull* dates back to the 1540s, none of the more than thirty gull species are actually named seagull. Besides, these birds were far inland.) This was a flock of California gulls, and they ate the crickets, saving the crops. The people of Utah were so grateful that they finally declared the species the official state bird in 1955. Technically the legislative code lists it as "sea gull," but most ornithological scholars agree that California gull is what they meant. The gulls are also honored with a monument on Temple Square in downtown Salt Lake City.

Like most gulls, California gulls are generalist feeders, equally as adept at taking down crickets as they are at dumpster diving. They breed from central California to northern Canada. Most of the species winters along the Pacific coasts, although some birds stay inland year-round. In Utah, you can spot California gulls in all seasons, especially near the Great Salt Lake.

ON TOP OF THE WORLD

Life at the top of the world is fairly similar around the globe. Circumpolar plants and animals found across the upper tier of North America, Europe, and Asia are adapted to cold extremes. Polar bears, walruses, Arctic fox, and musk ox eke out a living, be it in Greenland or Alaska. Birds, including the snowy owl, snow bunting, and king eider, also show this doughnut distribution around the North Pole. Circumpolar wildlife present unique conservation challenges. Increasing global temperatures have the biggest impacts at the highest latitudes. Circumpolar species may prove to be the next canaries in the coal mine, serving as early indicators of dire conditions.

Sometimes the same circumpolar species may go by different names, depending on the region. Boreal owls are the same as Tengmalm's owl. *Alces alces* is another species with two names: In North America this is a moose. In Eurasia the largest member of the deer family is called elk. These regional differences are akin to how US soccer is termed football in the majority of the world. Perhaps sports should have scientific names too.

Circumpolar Wildlife

JULY 26

THE RED CAPS ARE COMING

Lichens are composite organisms that are made up of a fungus and either an alga or a cyano-bacterium. Of the thousands of kinds of lichen, a few are quite noticeable. The crustose types of lichens appear to grow crust-like, often on rocks. Other lichens look more leaflike and are collectively thought of as the foliose lichens. Fruticose types of lichen appear branch-like.

In the case of the British soldier lichen, for example, each grayish-green stalk stands an inch or so in height and is topped with a bright-red knob. Reproductive structures called apothecia, these red caps reminded folks of caps worn by British soldiers, and the name stuck. The stalks grow in clusters and can often be found at the base of decaying logs but are sometimes in loose soil or on rocks or trees.

JULY 27

BLOWING FURIOUSLY

Epic dust storms occur in arid desert locations worldwide, including, most notably, in the south-western United States. Dust storms known as haboobs—the name is derived from the Arabic term for "blowing furiously"—are especially impressive. They can spring up seemingly out of nowhere and are formed as thunderstorm systems collapse. The stiff winds associated with the weather front create a downburst that kicks up dust on an incredible scale.

Haboobs can measure sixty miles wide, and the daunting leading edge of the dust storm can be a wall thousands of feet tall. The systems pass as quickly as they materialize, but conditions can be quite treacherous in the moment. Rains from the storm system can evaporate before reaching the ground, although sometimes precipitation does occur on the tail end of a haboob system.

JULY 28

FEATHER TYPES

Not all birds fly, but they all have feathers. It's one of the diagnostic features that separate them from the other classes of animals. Feathers come in a variety of forms and functions. Down feathers lie close to the skin and insulate the bird. Hatchling birds are either born with down or grow some in their first few days, but adults also possess these feathers. Similar in function to down, semiplumes have more structure. Contour, or body, feathers overlap like roof shingles and offers some protection from the elements. Many species have a uropygial, or preen, gland that releases an oil, which they spread when preening to waterproof the feathers. Wing feathers (remiges) are asymmetrical, with a leading edge that is shorter and stiffer. Feather shape also provides the lift needed for flight, and tail feathers (rectrices) are often arranged in a fan shape and help serve as a rudder while birds fly. Specialized bristle feathers look like hairs, but they can function like eyelashes; they are essentially the shaft, or rachis, of a feather. In flycatching birds the rictal bristles help funnel prey into the mouths and provide a wider target to feed with. Along with coloration, feathers can be important in displays, and they can also help with camouflage.

JULY 29
FLYING SPIDERS

You may not think of spiders as great fliers, but they are. These arachnids disperse using a gravity-cheating technique called ballooning. For the longest time it was thought that spiders simply used bits of silk to catch rides on the breeze, but research now indicates that spiders use electrostatic forces to help provide lift. Spiders generally lift off on days with the gentlest of breezes. They will test the airs with sensitive hairlike structures on their front legs. Then the spiders tiptoe up high on all eight legs. Next they point their abdomen and spinnerets sky high. Long strands of silk eventually extend from the body, and the spiders are swept up and away.

Ballooning is most often exhibited by young spiders traveling relatively short distances, but spiders have been documented more than three miles above Earth's surface and out at sea. Darwin witnessed spiders land on his ship when he was far from shore. And Walt Whitman gave ballooning a nod in his 1868 poem "A Noiseless Patient Spider," originally included as a stanza in *Whispers of Heavenly Death.*

Spider Ballooning

BEARGRASS

For short bursts of time, fields of beargrass blanket the subalpine slopes of Glacier National Park during the summer season. The timing is imprecise from year to year, but mid-July and early August are the most consistent producers. Higher elevation sites bloom later than lower locations. Beargrass, which looks a bit like a single end of a giant cotton swab, is not a true grass, but is in the lily family and grows from the Rockies westward, as well as in British Columbia and Alberta. Individual plants bloom only every five to seven years, and the petals are shed after a few fleeting days.

JULY 30, 1921

Allegany State Park is established in western New York near Jamestown, the hometown of famed naturalist and ornithologist Roger Tory Peterson (see August 6).

Members of the Lewis and Clark Expedition named it beargrass, but the species is also called elk grass, basket grass, and soap grass. The plant has a long history of being harvested, dried, and woven into baskets, hats, cups, and other goods. Beargrass is fire-resistant and grows back quickly from rhizomes after a burn.

CHANGING LANDSCAPES

The process by which ecosystems change over time, succession occurs in two forms: primary and secondary. Primary succession is the creation of entirely new habitat or habitat that has never been colonized before. The classic example is vulcanization; fresh lava is an ecological blank slate. Retreating glaciers, shifting sand dunes, and freshly fractured rock surfaces can also demonstrate primary succession. Secondary succession occurs following a change in the landscape, a fire or an extensive blowdown, for example.

Species are all adapted to different environmental conditions. Succession is easiest to visualize with plant communities. There are suites of plants that thrive on disturbed soils and under sunny conditions. These pioneer species are the first to colonize areas. As the site changes over time, the location will become suitable for new species. While the term *climax ecosystem* is applied to late successional conditions, implying the end of the process has been reached, succession is a dynamic and constantly shifting process. Although the species composition varies by location, the concept remains true throughout the world.

FEELING COMFORTABLE IN YOUR SKIN

Beluga Whale

A lot of animals go through seasonal changes in fur, growing thicker coats during winter and shedding some of their fur in spring. Most mammals, including humans, constantly shed a few flakes of skin cells too. But a few species of whales take skin-shedding to an extreme. They experience an annual sloughing of skin more akin to a snake molt. The orcas, belugas, and bowhead whales have all been documented experiencing a total replacement of skin cells. Researchers have even seen whales rubbing on rocky crags, using them as natural scratching posts to help facilitate exfoliation. Theories on why whales shed their skin include that the process may remove external parasites or sun-damaged skin. Another theory is that the new skin acts as a fresh wet suit and that a whale's hydrodynamic efficiency improves after it molts.

OYSTERCATCHERS

Hefty shorebirds, oystercatchers have stout orange bills perfect for prying open large mollusks, including clams and oysters. Black oystercatchers are found along the Pacific coast from Mexico to Alaska, while the American oystercatcher is the species on the Atlantic and Gulf coasts and along the Gulf of California in the Pacific. They often hybridize in Southern California and western Mexico, where the two species' ranges overlap. In July and August, oystercatchers may gather in loose flocks prior to migration.

AUGUST 2, 1956

Occupying most of Saint John, Virgin Islands National Park is created on this date. The area is recognized for coral reefs, tropical forests, and the ruins of an eighteenth-century sugar plantation.

Their migration patterns are fascinating. In one case, a pair of nestlings that hatched in North Carolina moved in opposite directions. One headed north to coastal Virginia, while the other moved to South Carolina. And northern birds often leapfrog over nonmigratory birds and winter farther to the south. Young birds will often spend multiple years in the South before moving north again for breeding. Many populations of oystercatchers are resident year-round.

AUGUST 3

LONG TIME NO SEE

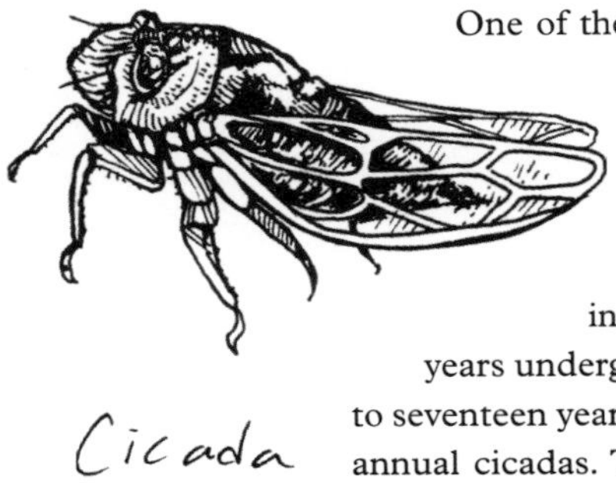

One of the soundtracks of summer in some regions is the raspy buzz of cicadas. These peanut-in-the-shell-sized insects make quite the commotion from the trees; it can approach 100 decibels. They make this sound by contracting an internal membrane, which is a bit drum-like, and their mostly hollow abdomen amplifies it. Periodic cicadas are cyclical. All of the individuals in a population are uniform in age, and they spend years underground as nymphs, with adults emerging all at the same time up to seventeen years apart. The majority of the more than 150 different species are annual cicadas. These types have staggered age classes, so individuals emerge every year. Cicadas slurp up plant xylem, and they rid themselves of excess liquids with little droplets of honeydew.

Some people call cicadas locusts, but that's a different type of insect entirely. Cicadas' scientific name means "tree cricket," but they aren't true crickets either. Though the hearty invertebrates look daunting, they are truly harmless to people.

AUGUST 4

DESERT CARDINAL

A close relative of the northern cardinal, the pyrrhuloxia, or desert cardinal, is found in southern Arizona, New Mexico, Texas, and central Mexico. The name pyrrhuloxia is a combination of the Latin term for bullfinch and a Greek reference to the shape of the beak on the bird. Both sexes of the pyrrhuloxia look similar to female cardinals, but the males display more red than the females. The pair is territorial during the breeding season, although the turf battles don't extend to cardinals that overlap a home range. In the wintertime, flocks of up to a thousand pyrrhuloxia gather.

The species uses its heavy yellow bill for seed cracking, and they also eat insects when available. Although they are more likely to feed on the ground than to come to a feeder, they can sometimes be enticed into backyards with sunflower seeds.

SQUIRRELS GONE WILD

The old saying goes that historically a squirrel could travel from the Atlantic Ocean to the Mississippi River and never touch the ground. Though tree cover isn't as extensive as it once was, squirrels are still quite widespread. Flying squirrels are highly arboreal. The so-called tree squirrels are quite comfortable toggling between the ground and the canopy, and curiously, a few of the ground squirrels are also noted for their climbing abilities.

During the coldest days of winter, natural tree cavities, or those excavated by woodpeckers, are often the preferred roost of tree squirrels. But on hot summer days, they take advantage of structures they build themselves, leaf roosts called dreys. They nibble and cut twigs with leaves still attached and weave them together to create cozy crannies. Individual squirrels build multiple dreys. These leaf houses tend to be used solitarily, although sometimes they are utilized as maternal sites, despite offering less protection from potential predators. The bundles of leaves are nearly impossible to see in the dense foliage, but the clumps are easier to spot in winter.

Squirrel Drey

FIELDS OF GUIDES

These days there are hundreds of natural history guides in print. Hundreds. And while Roger Tory Peterson's seminal *A Field Guide to the Birds* (1934) wasn't the first attempt at a field guide, it was unquestionably the one that upped the ante, launching the modern style we know today, with added emphasis placed on key identifiable field marks. The first printing of the book sold out within a week and has never stopped selling over its subsequent five editions, the most recent of which was published in 2008.

What makes this title exceptional isn't simply Peterson's expertise as an ornithologist, it was also his passion as a wildlife artist. In 1927 Peterson attended the Art Students League in New York City, followed by the National Academy of Design in 1929. His artful eye and clear illustrations beautifully depict identifiable field marks, inspiring the public to take interest in identifying birds. His life's artwork remains on exhibit at the Roger Tory Peterson Institute of Natural History in his hometown of Jamestown, New York.

HORNY TOADS

Horny toads, as they are affectionately termed, are technically called short-horned lizards. Taxonomic changes continue to reclassify the family, but fifteen or so species are found north of Mexico. Characteristically they share short squat legs, rounded bodies that are pancake flat, and spiky blocky heads, and they all rely heavily on camouflage to blend into their arid habitats. These sit-and-wait feeders predominately target ants but will opportunistically take other insects. When attacked, especially by canids, most species of horny toads have the curious ability to squirt blood from their eye sockets, an effective avoidance strategy.

Greater Short-horned Lizard

These lizards are one of the few reptiles who are ovoviviparous, meaning that they give birth to live young. Females will deliver anywhere from five to nearly fifty 1-gram babies between July and September. Two states recognize horny toads as their state reptiles: the greater short-horned lizard in Wyoming and the Texas horned lizard in Texas.

INLAND SEA

The Great Salt Lake is a remnant of Lake Bonneville, an inland sea that covered twenty thousand square miles around twenty thousand years ago. The water covered much of what is now Utah and parts of adjacent Idaho and Nevada. Today it's the largest saltwater lake in the Western Hemisphere and the biggest body of water between the Great Lakes and the Pacific Ocean. Its salinity levels fluctuate depending on water levels and runoff but are often well above that of the oceans. Brine shrimp is one of the few organisms that thrive under these conditions. These aquatic invertebrates can go through four or more generations in a single growing season. The shrimp feed on algae and in turn are a major source of food for millions of migratory birds each year.

Cinnamon teal, black-necked stilts, American avocet, and white-faced ibis all utilize Great Salt Lake in huge numbers. It is estimated that 50 percent of the world's population of Wilson's phalaropes spend time at the Great Salt Lake during fall migration. Similarly, eared grebes congregate at Great Salt Lake (along with California's Mono Lake) in the fall.

DISCOVERING DARTERS

If darters were a bit bigger, they'd be the most well-known fishes around based on their stunning coloration alone. One species of darter is actually called rainbow darter, but many species are brightly patterned. There are more than 225 species of darter, which are in the perch family, and more are being discovered yearly. They are important to their native ecosystems, and the southeastern United States has the greatest diversity of darters. At just two to three inches, they are easy to overlook, though.

One species, the snail darter, captured the attention of the Supreme Court in the 1970s, in an early test of the Endangered Species Act. The construction of the Tellico Dam by the Tennessee Valley Authority would have likely led to the extinction of the small fish. The court temporarily halted the dam project until managers could protect the species, but Congress eventually passed an exemption for the Tellico project. Snail darters were relocated to other rivers (which it was later discovered impacted another endangered darter, the sharphead), and the dam was finished in 1979. Since then, additional populations of snail darters have been discovered, so the species was downlisted from endangered to threatened. As for other species of darters, they face the common pressures of polluted water, altered stream flows, and competition from nonnative species. Restoration efforts are under way to support darter species throughout the Southeast.

MINERS NOT WEASELS

Badgers' reputations often precede them. People rarely encounter them, but it seems that nearly everyone thinks badgers are ferocious. When cornered, badgers have an intimidating snarl, but unless you are a rodent, you're likely safe from this maligned mammal. These stocky weasels are found from open deserts to mountain meadows but are especially at home on the prairies. Most observations of badgers come in fall, when the normally solitary males seek out mates. Due to delayed implantation, young badgers aren't born until the following spring. Long claws let badgers dig with ease, and they feed extensively on rodents, from mice to prairie dogs.

Wisconsin, the Badger State, is home to relatively few badgers. Instead, the nickname is a tribute to 1800s miners who also spent considerable time tunneling underground as they extracted what became Wisconsin's state mineral, galena, an important source of lead and silver.

IGNEOUS ROCKS

Igneous rocks form when molten magma from the core or mantle of the earth cools and hardens, which can happen in pockets beneath the surface or after lava erupts from volcanoes. The basalts are igneous, and Craters of the Moon National Monument and Preserve in Idaho protects many varieties of basaltic lava rocks. Obsidian also has volcanic origins. Dark, shiny, and glass-like, obsidian is formed when lava cools rapidly—crystals do not develop before the rock hardens. A lack of geologically recent volcanic activity means obsidian is not found east of the Mississippi River. By contrast, California has a number of obsidian deposits, especially at Long Valley Caldera and Mono-Inyo Craters. The rock was knapped into knives, blades, and spear or arrow tips. It is also popular in jewelry. Igneous is one of the three types of rocks, along with sedimentary (see December 2) and metamorphic (see June 8).

PERSEID METEOR SHOWER

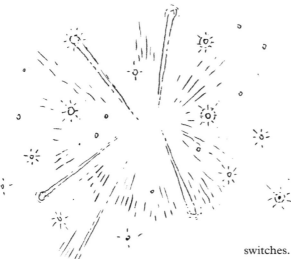

perseids

AUGUST 12, 1971

A story exploring environmental degradation and human responsibility, *The Lorax* by Dr. Seuss is published. Decades later, the book continues to remind readers that "unless someone like you cares a whole awful lot, nothing is going to get better. It's not."

Skip the late-night television reruns tonight and head outside. Because in mid-August the Perseid meteor shower peaks! Don't panic if it's cloudy; meteor showers don't turn on and off like light switches. You can spot these meteors for a few days before and after the peak. Viewing is best after midnight and into the early-morning hours. Getting away from city lights is also helpful. Look for the meteors in the northeast sky, near the constellation Perseus.

Meteors aren't falling stars. Meteor showers, like this week's Perseids, happen when Earth passes through the tail of a comet or asteroid, in this case, the Swift-Tuttle Comet. The meteors, which can fall as fast as 45,000 miles per hour, are rocky or metallic pieces that burn up as they enter Earth's atmosphere. When a meteor strikes Earth's surface, it is known as a meteorite.

DOUBLE-DIPPING

The ability to swim and dive is a given for ducks and geese but a more unexpected feat for a song-bird. Cue one of John Muir's favorites, the American dipper. Formerly known as the water ouzel, these slate-gray passerines are the size of portly potatoes. Dippers thrive in fast-moving headwa-ters and rushing streams from Alaska to Central America. They walk against the current along the bottom of these waters, foraging for invertebrate larvae, small fish, and fish eggs. Dippers are cold tolerant, thanks in part to dense feathers and a low metabolic rate.

Unusual for songbirds, they undergo a molt of wing and tail feathers in summer, which leaves the birds flightless for a short period of time. Streamside nesting locations, secure from predators and protected from flooding, appear to limit their distribution. Dippers prefer ledges or crevices above deeper stream pools. The birds weave a mossy outer shell up to ten inches in diameter to protect an inner cup nest from moisture.

THE GRASSLANDS

Grasslands are a dominant biome found on six of the seven continents. The United States has three major clas-sifications of grassland prairies: shortgrass, mixed-grass, and tallgrass. Shortgrass prairies are common to the Great Plains from central Alberta to eastern New Mexico. These areas are in the rain shadow of the Rockies, where precipita-tion is lower. Buffalo grass and blue grama are the dominant species. To the east, the mixed-grass prairie is a transition zone. This swath of habitat runs through the midsection of Saskatchewan, the Dakotas, Nebraska, Kansas, Oklahoma, and into Texas. Here species are varied but can include the

Tallgrass Prairie, Coneflower

bluestems, western wheatgrass, and a number of forbs. Tallgrass prairies are farther east, from Man-itoba to East Texas. Here productive soils and increased precipitation, closer to forty inches per year, support plant species that grow six to ten feet high, including bluestem, coneflower, and compass plant (see October 29). The Forest Service administers over a dozen national grasslands, great places for exploring your public lands.

AUGUST 15

SPREAD 'EM

Have you ever seen a perched bird with its wings splayed out to the sides? It certainly looks uncomfortable, but it's got to serve a purpose, right? If you're looking at a waterbird flashing their fully extended wings, chances are it's either a cormorant or perhaps an anhinga if you're in southern Gulf states. These divers look duck-like, but they aren't related at all. Cormorants and anhingas lack the ability to fully waterproof their feathers like waterfowl. Instead, after pursuing an underwater meal of fish, these species must dry out their feathers, usually from a prominent branch exposed to the sun.

Another family of birds, the vultures, is also well known for basking in the sun. The birds extend their wings to help with thermoregulation and to potentially kill off bacteria. Catching the rays can drive up their temperatures, so vultures also have another solution to prevent overheating. They defecate on their legs for the evaporative cooling effect.

AUGUST 16

ENDANGERED RECOVERIES

Sometimes it can feel like conservation causes are filled with doom-and-gloom stories. Although many species are still in peril, there have been plenty of ecological recoveries. Some of the most common species of today were at one time nearly wiped out, and often less than 150 years ago. By the mid-1880s, market hunting and habitat conversion were already taking a toll on wildlife, especially big game like bison, elk, and pronghorn. Fashion trends, such as beaver hats and egret feather–plume adornments, also drove the exploitation of nature. Conservation laws were passed, many organizations were founded to champion the causes of wildlife, and animal populations slowly recovered. In the mid-1900s, however, new threats emerged. Pesticide use was linked to thinning eggshells for bird species like bald eagle, osprey, and brown pelican. Again, people rallied behind the conservation cause.

More recently, the lesser long-nosed bat, a species that pollinates agave plants, was nearly wiped out. Collaborations between the US and Mexico have brought it back from the brink of extinction. It was the first bat species to be delisted, or removed from the list of species deemed endangered or threatened under the Endangered Species Act. So raise a shot glass of tequila for the bats and all the other conservation success stories. Cheers.

BEEP-BEEP

Scampering about in the Southwest, greater roadrunners are large cuckoos. The state birds of New Mexico, they've expanded their range north to central California and east to Louisiana and Missouri. Like other species in the family, roadrunners have zygodactyl feet, with two toes pointing forward and two in the back (see November 21). The distinctive X-shaped track is a sacred symbol to Pueblo tribes. According to legend, the symmetrical track disguises the direction the bird is traveling, therefore preventing evil spirits from following.

In the cartoons, Road Runner always avoided Wile E. Coyote, but in reality the mammalian predators are a threat to the birds. In turn, roadrunners are apt predators of an array of prey, including numerous venomous species such as lizards, scorpions, and rattlesnakes. They swallow long snakes bit by bit as they digest them. Roadrunners get enough moisture from their food, so they never need to drink water. Excess salts are discarded from glands near the birds' eyes.

Greater Roadrunner

HOT WATER

When the temperature of water rises, the liquid contains less dissolved oxygen. Some species tolerate these seasonal fluctuations, but cold-water specialists, like trout, are stressed by warmer summer temperatures. Fifty degrees Fahrenheit is comfortable for trout. When water temperatures reach 65 degrees Fahrenheit, fish can become lethargic and slower to feed, and extended exposure to warmer conditions can stunt their growth. As water temperatures exceed 70 degrees Fahrenheit, trout can become heat-stressed.

Summer anglers are encouraged to limit fishing when conditions get extreme as studies show that mortality rates increase. To keep it cool, head for higher elevations, spring-fed streams, or deeper bodies of water. Fishing in the early hours can also be productive and safer for the trout.

FIREWEED FIRECRACKERS

Fireweed is a classic early successional plant species (see July 31). It grows across much of the continent and thrives in areas of recent disturbance. It was one of the first species documented following the eruption of Washington's Mount St. Helens. As the name implies, this species is often found following a forest fire but also does well in avalanche chutes and even roadside ditches. The green stems stand tall, and beautiful magenta-colored blossoms lined up in rows offer a vibrant sign of life in a sometimes-stark, disturbed landscape.

Thanks to regeneration from rhizomes and an abundance of wind-dispersed seeds, fireweed can move into an area quickly. Peak abundance occurs anywhere from two or three years to seven or eight years after the plant has become established. A decline in fireweed is often attributed to shade intolerance as shrubs and tree saplings slowly take over terrain.

Fireweed syrup, jelly, and honey are popular in Alaska. According to Alaska Native lore, fireweed is seen as a marker for the seasons. When the tops of fireweed are in bloom, the first snows are said to be about six weeks away.

THIN AS A RAIL

The saying "thin as a rail" is widespread, especially in bird circles, but the ornithological connection is a stretch. The phrase dates to the late 1800s. Since variations include "thin as a lath" and "thin as a rake," the *rail* in question was surely of the wooden variety initially and not a reference to the wading birds known as rails. Birds in the rail family do tend to be laterally compressed, so the skinny reference holds true. Rails thrive by slinking through dense marsh vegetation. Although they can be difficult to spot, distinctive clicking calls reveal rails.

Sora is the most abundant and widespread of the North American rails, followed by the Virginia rail. Ridgway's rail was classified as a new species separate from the clapper rail in 2014 and can be found in the Southwest. Clappers are the rail of the salt marshes of the Atlantic and Gulf coasts. Similar in appearance to clappers, king rails live in freshwater marshes of the Southeast. As reclusive as they are thin, black and yellow rails are two of the most difficult bird species to spot.

King Rail

HOWLING AT THE MOON

Seeds and vegetation are staples in the diets of most rodents, but not grasshopper mice. These most-impressive carnivores live on a diet of insects, lizards, and even other mice and voles. Grasshopper mice will cache food for lean times. Southern grasshopper mice specialize on another unexpected prey choice—scorpions—because the mice are immune to their venom. Instead of experiencing the pain of the toxin, neurologically, the mice are able to use the venom as a painkiller.

Another characteristic of this mouse reminds people of larger carnivores—namely wolves and coyotes. The highly territorial micropredators communicate over distances by emitting loud howls. Relatives of the genus *Onychomys* found in the fossil record have been dated back to the early Pliocene, 5.3 million to 3.6 million years ago.

Grasshopper Mouse

HURRICANE NAME GAMES

The season for hurricanes runs from May 15 to November 30 in the eastern Pacific and kicks off a couple of weeks later in the Atlantic basin. Mid-August to late October tends to be the peak storm season, though strong storms can occur anytime. The practice of naming storms dates back to at least the late 1800s, but it has come and gone over the years. The World Meteorological Organization has an established protocol for identifying hurricanes: For Atlantic basin storms, six lists of names rotate over the years. Names are rolled out alphabetically each season, and names of particularly strong storms are retired.

The strength of a hurricane is based on sustained wind speeds and the Saffir-Simpson scale, which was developed in 1971 by Herbert Saffir, a structural civil engineer, and Robert Simpson, then the director of the National Hurricane Center in the United States. Tropical storms get upgraded to Category 1 hurricanes when they produce sustained winds of at least 74 miles per hour for one full minute. Category 5 storms have sustained winds above 156 miles per hour.

SHIFTING LINES

Nearly ten years after famously leading the 1869 expedition down the Grand Canyon, geologist John Wesley Powell noted an implied climatic bifurcation in North America. Later, in 1890, he wrote about this dividing line: "Passing from east to west across this belt a wonderful transformation is observed." He observed a "luxuriant growth" of grass and gaudy flowers on the east of the divide, while to the west, the ground "gradually becomes naked," with sparse bunch grasses, thorny cactus, and the sharp bayonets of yucca. The split runs roughly down the 100th meridian of longitude (from the Dakotas through Texas). Or at least it did during Powell's time.

The climatic patterns have shifted this drier zone east by 140 miles or so, closer to the 98th meridian these days. The rain shadow of the Rockies is partially responsible for the drier climates that stretch along the Front Range. Moisture from the Gulf adds precipitation east of this line. These patterns explain why wheat is the crop of the West, while the Midwest can support the thirstier crops of corn and soybeans. The eastward extension of the drier climate is attributed to warmer temperatures increasing evaporation in the North and less precipitation overall in the South.

FAIRY RINGS

A fairy ring is not a tiny silver band that slips around the finger of a fairy but a mushroom that sometimes pops up in backyards and neighborhood parks. They start out in a single spot, but as the fungus expands outward, an arc or even a full circle of mushrooms can grow. Some cultures have thought fairy rings were bad omens of things to come, while others have celebrated these curious fungal features.

Around sixty species of mushrooms will grow in ringed clusters, in both forested and meadow habitats. Forested species tend to be mycorrhizal (meaning symbiotic with a host plant's root system, such as tree roots), and meadow types are usually saprotrophic decomposers, which obtain nutrients from nonliving organic matter. As the nutrients are depleted, the fungus continues to expand outward.

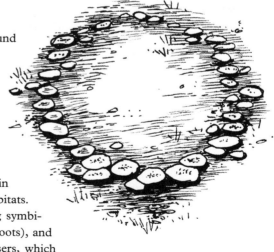

Fairy Ring

THE FALL MIGRATION OF SUMMER

Shortly after the latest-arriving songbirds wrap up their spring migration, the shorebird like the sandpipers, plovers, and others begin to head south. Many species of shorebirds nest in the high Arctic, and the first "fall" migrants can move through the northern United States by late June, and by July migration is in full swing. Adult males, likely the earliest migrants, often hit the skies shortly after breeding, sometimes

AUGUST 25, 1916

President Woodrow Wilson creates the National Park Service, within the Department of the Interior, to coordinate the management of America's federal parklands.

before eggs have hatched. Look for the yellowlegs species and least sandpipers first. Adult females follow behind. Juvenile birds are left to fend for themselves shortly after hatching. These young shorebirds move south in great numbers up to a month after the adults.

Late August and early September is perhaps the greatest shorebird-watching season since adults and juveniles of many species can be found. The latest of the shorebirds may still be moving south in December, and cold-hardy species like Wilson's snipe can even overwinter in the northern states and southern Canada.

FROM SHORE TO SHINING SHORE

Though there are some exceptions (looking at you, woodcock), most shorebirds are built for long-distance travel. They have sleek bodies and long pointed wings for added efficiency. Nearly twenty species make nonstop flights of more than 3,100 miles. The godwit takes this journey to the extreme. A bar-tailed godwit outfitted with a satellite tracker was recorded making a nonstop flight from the coast of Alaska to New Zealand. The journey took nine days to complete, and researchers calculated that the bird's average speed was thirty-five miles per hour over the 7,270-mile jaunt.

Birds don't usually have any body fat, but before a big migration, their bodies store fat to use. Godwits have been documented doubling in body weight in the days leading up to the migration. Their pectoral flight muscles increased in size, while their stomachs and gizzards were reduced in size. Upon arrival in New Zealand, the birds had lost over half their body weight.

LEWS

Northern watersnakes are fairly common, but one population wasn't faring as well in the 1990s. The Lake Erie watersnake (LEWS), found in the western basin of the lake, was listed as threatened in 1999. For over a decade, an extensive education and outreach campaign championed the cause of these island dwellers, and shoreline habitat protections were implemented, leading to the species being downlisted by 2011. These nonvenomous snakes feed extensively on aquatic prey. Historically the catch was heavy on amphibians and fish, including madtom, stonecat, log perch, and spottailed shiners. In the 1990s, the nonnative round goby led to population crashes of many of these native species. These days, LEWS get up to 90 percent of their calories from goby.

The snakes are ovoviviparous, so after breeding in the spring, the females carry eggs that hatch inside their bodies in August or September. They then give birth to live young.

BORDERLAND BUTTERFLIES

The Lower Rio Grande Valley of Texas is at the intersection of tropical and temperate climates. The region has the longest growing season in the United States, with average temperatures of 72 degrees Fahrenheit and 325 days of sun annually. These conditions make for unique ecology. With nearly 150 species of butterfly that can be found only in the Lower Rio Grande Valley and points more southerly, the border has impressive insect diversity. The region hosts nearly 40 percent of the more than seven hundred butterfly species found in the United States. Species like malachites, Julia's, red rims, and Mexican bluewings are stunning. In an area now dominated by agriculture, small pockets of wildlands remain in state parks and private preserves, including the National Butterfly Center. The butterfly gardens at these preserves are hot spots for other pollinators like native bees too. The region is also home to a remarkable diversity of bird species (see November 9).

ANCIENT WING

Archaeopteryx (meaning "ancient wing") is thought to be the original bird with characteristics of both small, featherless Mesozoic dinosaurs and modern birds. Raven-sized, the species likely moved through the air, although there are ongoing debates as to whether it was a glider or an active flapper. The dinosaur had small teeth, clawed fingers, and a long bony tail. *Archaeopteryx* fossils date to the late-Jurassic period, approximately 150 million years ago. The type specimen was first described in 1861, just a couple of years after Darwin published *On the Origin of Species*. Since then, ten more *Archaeopteryx* specimens have been discovered in the Bavarian region of Germany. Additionally, specimens of *Anchiornis*, *Xiaotingia*, and *Aurornis* have been unearthed, all of which represent transitional fossils in the earliest development of feathers.

Archaeopteryx

CONVERGENT EVOLUTION

With convergent evolution, two or more unrelated species develop similar characteristics to serve the same ecological niche functions. Hummingbirds are new-world birds. Africa doesn't have hummingbirds, but the continent is home to a number of doppelgängers known as sunbirds. Both complexes of birds have long thin bills for probing flower blossoms, and

AUGUST 30

International Whale Shark Day is a celebration of the world's largest fish, which can reach nearly forty-two feet in length and weigh about 47,000 pounds.

both hummingbirds and sunbirds are critical pollinators for their local flora. They look like kin, but sunbirds can't hover and fly backward like hummingbirds. Sunbirds are more similar genetically to crows, and hummingbirds' closest relatives are the swifts.

Another unrelated critter mistaken for a hummingbird in backyards across the United States is the hummingbird moth, which hovers at flowers as it feeds. Active during the day, the rapid-fluttering moths are hummingbird doppelgängers in form and function.

THE HIBERNATION DAYS OF SUMMER

On the hottest days, a few animals can enter a summer hibernation of sorts, either a daily torpor or longer bouts called estivation. The desert tortoise is one species that will remain inactive for a few days. And some of the hibernators, especially the rodents, settle in for their long winter of true hibernation by late summer. Richardson's ground squirrels inhabit the prairie provinces of Canada,

Montana, Idaho, North Dakota, South Dakota, and western Minnesota. For this species, the only individuals active in late August are likely that year's young. Adult males can begin hibernation by Summer Solstice, and females can enter the physiological state in July.

SEPTEMBER 1

WHEN DAY TURNS TO NIGHT

A solar eclipse occurs when the moon passes between the sun and Earth. While solar eclipses are rare for any location, they are more common than you may think. On average, locations have to wait roughly 360 years between total solar eclipses, but one happens every eighteen months or so somewhere on Earth. Adding to the perceived rarity is the fact that just a narrow band of viewers can see the full phenomenon, when the sun figuratively sets. The shadow covers a mere 166-mile wide swath. And remember that more than 70 percent of the planet surface is ocean, so your eclipse viewing options on land are limited.

If you are lucky enough to live in the path of totality, be prepared for some of the most memorable 450 seconds of your life. If you don't live in the path, a seven-and-a-half-minute wonder may not seem worth the effort, but it *totality* is.

SEPTEMBER 2

UNDERWATER CATS

Like the whiskers on your feline friends, the facial appendages of catfish serve a sensory function. These fleshy whiskers, or more technically barbels, help the fish gather scent and taste information. Basically the entire body of a catfish is one big taste bud. They're covered with more than one hundred thousand of these sensors. Some folks fear that the whiskers are stingers. That's not the case, though catfish do have sharp spines on their pectoral and dorsal fins that can stab you.

SEPTEMBER 2, 1937

President Franklin Delano Roosevelt signs the Pittman-Robertson Act to fund wildlife conservation and hunter education efforts with revenues generated from excise taxes on hunting equipment sales. The Dingell-Johnson Act extends a similar program for fisheries in 1950.

There are numerous species of catfish, but based on state designations alone, channel catfish seems to be the most popular. This species is recognized by five states: Iowa, Kansas, Missouri, Nebraska, and Tennessee. Catfish are popular with anglers, who may use stink baits and chicken livers as traditional offerings, although a few folks prefer the noodling technique of catching catfish with their bare hands.

UNDERGROUND OWLS

Whether it's on the prairie dog towns of the West or the South Florida golf courses, the burrowing owl is the exception to many of the owl rules. For one, these eight-inch leggy munchkins are more diurnal than most owls. And unlike other owls (and the majority of birds of prey), both sexes of burrowing owls are similar in size: They don't show the usual reverse sexual dimorphism of raptors, where females are larger than males. Plenty of species of owls utilize cavities in trees, cacti, or manmade boxes or barns, but the burrowing owl lives a partially subterranean existence. They live in loose colonies, perhaps due to their specialized habitat requirements. Burrowing owls will dig their own burrows in the sandy soils of Florida, but they rely on existing tunnels in other habitats. They are more tolerant of carbon dioxide than other birds, an adaptation that helps them survive for extended periods underground. Some owls will collect animal dung as a way to chum in insects for easy consumption. Although southwestern and Florida birds are year-round residents, northern populations of burrowing owls in the West migrate south, another trait shared by few other owls.

Burrowing Owl

NATIVE THISTLE

All thistles often get lumped together as bad invasive species, but in fact, there are a number of native species. Due in part to the expansion of nonnative types, five species of *Cirsium* thistles are classified as endangered in the United States. Localized species, like Pitcher's thistle, are in peril because of habitat loss and degradation. In the Midwest, native thistles support at least eleven species of bumblebees, including the endangered rusty-patched variety. Illinois has six species of native thistle that are pollinated by at least eighteen different kinds of butterflies. And it's not just the insects that benefit from the thistles; American goldfinches are late-nesting songbirds that utilize thistle for nesting material and as a food source. Many varieties of thistle are fuzzy with prickly spines, and others are nearly hairless; many parts of the plant, including the stalks, leaves, and flowers, are edible.

Elk thistle, a species found from Wyoming to Alaska, is sometimes called Evert's thistle. This name is a reference to Truman Everts, who was a member of an early expedition exploring the area that would eventually become Yellowstone. Everts became separated from the group and survived on not much more than the thistle for over a month before being rescued.

The thistle in your birdfeeder isn't the same as the type that grows wild. Instead this crop is from a different plant native to Ethiopia. The imported birdseed is heat-treated to prevent it from germinating.

▶ SEPTEMBER 5 ◀
ALLIGATOR NESTS

Alligator Nest

American alligators range along the southeastern coastal states from North Carolina to Texas and inland to Oklahoma and Arkansas. Gators are the largest reptiles in North America: Males grow to eleven feet on average and tip the scales at nearly one thousand pounds. During breeding season, alligators are quite territorial, with bellowing bulls announcing their presence. Their elaborate courtship involves vocalizations, pheromones, snout and back rubbing, body posturing, and water displays that include head-splashing and bubble-blowing. Females build large nests from mud and vegetation from April in South Florida to late June or July in the more northern populations. Average clutch size is about forty eggs, and females actively defend the nest for the sixty-five-day incubation period.

Alligators exhibit temperature-dependent sex determination: Eggs that incubate at cooler temperatures and hotter temperatures will develop as females, while intermediate temperatures lead to males. This thermal sensitivity occurs during the early weeks of incubation. Alligator hatchlings are six to eight inches long and highly susceptible to predation, with upward of 70 percent falling victim to birds, raccoons, bobcats, snakes, fish, or other gators.

▶ SEPTEMBER 6 ◀
FIN-FOOTED

Pinnipeds are the fin-footed mammals, including walruses, seals, and sea lions. These carnivores of the sea haul out on land or ice for resting and rearing pups. Odobenidae is the family of the walruses (see May 5). True seals, in the Phocidae family, have small front flippers and lack external ear flaps (see December 1). Members of the Otariidae family are known as eared seals, or fur seals and sea lions. The Otariids can walk on all four flippers and have noticeable earflaps. Fur seals can have more than three hundred thousand hairs per square inch of skin, a fact that nearly wiped many species out when furs were in high demand.

Take a stroll out to Sea Lion Point in California's Point Lobos State Natural Reserve for a memorable pinniped experience. The *lobos* are sea wolves, or sea lions. The barks of California sea lions can be heard from quite a distance. From mid-December to the end of March, California's Año Nuevo State Park offers guided hikes to see northern elephant seals.

SEPTEMBER 7

NUTTALL THE NATURALIST

Thomas Nuttall was an English-born botanist and zoologist who explored throughout America in the early 1800s. Nuttall described many plant species on an 1811 trip up the Missouri River, expanding on the natural history work of Lewis and Clark's earlier expedition. He also explored the Great Lakes region and the Louisiana Territory along the Arkansas and Red Rivers. By 1818 he published *The Genera of North American Plants*. In 1823, Nuttall was elected as Associate Fellow of the American Academy of Arts and Sciences, and two years later he was named curator of Harvard University's botanical gardens. Nuttall also had an interest in birds, publishing *A Manual of the Ornithology of the United States* in 1832 and a Canadian version a couple of years later. An 1834 expedition took him from Kansas to the Northwest and down the Snake and Columbia Rivers, and eventually to Hawai'i. Nuttall returned to the Pacific Northwest the following year, continuing to examine the plants he came across.

From 1836 to 1841 he worked for the Academy of Natural Sciences in Philadelphia, and despite some scientific squabbling about authorship, much of his fieldwork found its way into *A Flora of North America* by Asa Gray and John Torrey. Nuttall is credited for naming hundreds of species of plants. Additionally, a number of species have been named on his behalf, including Nuttall's violet, Nuttall's woodpecker, yellow-billed magpie (*Pica nuttalli*), and mountain cottontail (*Sylvilagus nuttallii*).

SEPTEMBER 8

BACKYARD HABITAT IS WHERE IT'S AT

Habitat doesn't have to be expansive swaths of untouched land. Postage-stamp-sized backyards can provide the essentials of food, water, shelter, and space. To attract wildlife to your neighborhood, start with a foundation of high-quality native plants. These local species are the key to building a backyard food chain. Invertebrates require specific plants to meet their needs, and ornamental plantings don't cut it. One study calculated that if a yard was made up of more than 30 percent nonnative plants, for example, chickadees wouldn't move in.

Another way to increase the wildlife value of your real estate is to offer sources of water. A simple birdbath helps, but be sure to change the water and clean the bowl regularly. Flowing water and natural-looking pondscapes are great too. Shelter in the form of shrubs, a brush pile, or native groundcover plantings can help protect wildlife from weather and predators. The National Wildlife Federation has a Certified Wildlife Habitat program to recognize high-quality spaces. With a little effort, you can welcome wildlife to your own backyard.

SEPTEMBER 9

WALKING WITH GIANTS

Fossils are fascinating to examine, but in situ dinosaur tracks really capture our imaginations. Most preserved dinosaur tracks were formed in wet mud or sand. A late-summer visit to the Ballroom site of Dinosaur Valley State Park in Glen Rose, Texas, can reveal prints normally covered up by a flowing stream. These prints were left behind by a number of sauropods, including adults and juveniles, when the location was at the edge of a giant ancient ocean. The story the tracks tell is a familiar one of a predator in pursuit of prey—a young sauropod was seemingly being pursued by a theropod. Elongated tracks and dewclaw marks show evidence of slippery conditions.

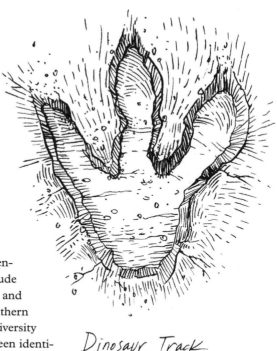

Dinosaur Track

Another location, Dinosaur Ridge, west of Denver, Colorado, has well-preserved tracks that include those of stegosaurus, apatosaurus, diplodocus, and allosaurus. Dinosaur Provincial Park in rural southern Alberta, Canada, has tracks and an impressive diversity of dinosaur fossils; more than fifty species have been identified in the region. New Mexico, Connecticut, and Massachusetts also have well-known locations for stepping into the past.

MISTAKEN IDENTITY

Environmental mimicry can serve different functions, such as to protect a creature or to lure in prey. Some species, like viceroy butterflies, mimic unpalatable look-alikes, including monarchs. Other populations of viceroy look more similar to the local species of queen butterflies and soldier butterflies. Giant swallowtails and other butterfly species take this specialized camouflage to the extreme as their caterpillars resemble bird droppings.

In a twist on the mimicry functions, some species of predatory fireflies mimic the flashing patterns of other species. Males are drawn in to what they perceive to be mating flashes from females of their species. Instead they are met with a predatory species luring them in with the hopes of scoring an easy meal.

In a few species of fish, males can look like females. These sneaker males are able to breed with females, unbeknownst to the dominant male. Even though species tend to look similar in mimicry, the phenomenon is different from convergent evolution (see August 30).

LIGHTS OUT

The majority of songbirds migrate at night, which makes them susceptible to light pollution from a growing human population and expanding urban footprint (see April 1). Each year, a powerful September 11 remembrance, the Tribute in Light memorial, shines two towers of light skyward for four miles from dusk to dawn on the anniversary of those tragic terrorist attacks. Because the date overlaps with the timing of prime fall migration, hundreds of birds can get disoriented by the glow emanating from Manhattan. Officials recognized this conflict, and so the lights are regularly dimmed to allow birds to disperse, based on criteria developed by New York City Audubon. Audubon volunteers monitor for the thresholds they've agreed to each year, and the lights are shut off when they reach that number to allow the birds to disperse. This same kind of solution can easily be applied to other forms of artificial lighting, including large buildings, expansive parking lots, and sports stadiums. Minimizing lighting reduces energy usage and prevents birds from getting disoriented by the lights and dying either of exhaustion or by colliding into buildings or other objects.

Light pollution isn't just problematic for birds. Sea turtles sometimes get turned around by lights near the beaches they nest on, and many insects are impacted by the increase in lumens, often congregating in the artificial brightness. Predator-prey dynamics can be shifted by the shine too. People have fundamentally altered the landscape (and the skyscape) through lighting, and we are increasingly examining the effects of these changes.

GHOST ORCHID

Found in Florida, Cuba, and the West Indies, the ghost orchid is a rare beauty that's endangered due to illegal collection, habitat loss, and changing wetland hydrology. Audubon's Corkscrew Swamp Sanctuary, in southwest Florida, has a ghost orchid, the largest discovered worldwide so far, visible from their boardwalk. The orchid can bloom any month of the year, but peak blooming occurs from July through September. The plant is leafless; it's chlorophyllous gray-green roots cling in the canopy of any number of tree species.

The ghost orchid is thought to be pollinated exclusively by the giant sphinx moth. This mammoth moth has a five- to seven-inch wingspan and a proboscis that is long enough to probe the lengthy nectar spur of the flower. Beyond Florida, the moth is found along the southern tier states and throughout Central America.

SEPTEMBER 12, 1964

Utah's Canyonlands National Park is created by President Lyndon B. Johnson. Distinct sections of the park include Island in the Sky, The Needles, and The Maze, along with the Colorado and Green Rivers and their tributaries.

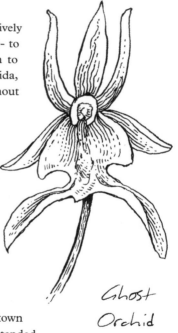

Ghost
Orchid

PRAIRIE PUPPIES

Prairie dogs are hefty ground squirrels found in the grasslands of the West. Of the five species of prairie dogs, black-tailed has the most extensive distribution, stretching from southern Canada to northern Mexico. Within this range, the species occupies only about 2 percent of its historic habitat due to habitat loss and eradication efforts. Prairie dogs are highly colonial, and a typical prairie dog town can stretch for acres. Within a given prairie dog town, extended family units tend to congregate in subunits known as coteries.

The genus, *Cynomys*, means "dog mouse," and the barks of prairie dogs inspired their name. The jump-yip display call seems to be a territorial communication. Prairie dogs also utter various alert calls when they spot predators. These warnings send the animals scurrying to nearby burrows. The endangered black-footed ferret (see May 17) and many other animals depend on this keystone species.

LOOKING FOR LOVE

A very different burrower, the tarantula, is found throughout the southern and southwestern United States. Urticating hairs on their legs and abdomens help the spiders detect vibrations. These hairs can cause some minor irritations to humans, but tarantulas are only slightly venomous, hesitant to bite, and pose little risk to people. These fascinating arachnids generally prey on insects, especially grasshoppers, beetles, and spiders. A single cricket can sustain a tarantula for a couple of weeks, so the spiders spend much of their time underground, avoiding hot summer and cold winter temperatures.

Tarantulas are primarily nocturnal, and the best time of year to spot them is evenings in the fall when males make big movements across the landscape. It's not a migration per se. Instead the sexually mature male spiders are seeking out females. The lady spiders are homebodies who can reproduce multiple times throughout their 30-year lifetimes. This time of year, the 7- to 12-year-old males travel up to fifty miles, following the female's pheromones, to make a match. Predators, including lizards, birds, and coyotes, will make a meal of these vagabonds moving across the landscape. Males are also sometimes killed by female spiders, either before or after mating. Even if they survive their conspecifics (member of the same species), they won't survive beyond the onset of winter.

SCISSORBILLS

Not even the founder of modern taxonomy, Carl Linnaeus, was immune from typos. The genus *Rynchops*, should be *Rhynchops*, from the Greek for "beak face." Alas, the misspelling lives on as the genera of the three species of skimmer: black, Indian, and African.

Black skimmer, the species of North America, can be found along the Atlantic and Gulf coasts and in the Pacific Ocean north to Central California. The tern-like species are the only birds whose lower beaks are longer than their upper ones.

Skimmers hatch with even mandibles, but within a month their upper and lower bills are noticeably different lengths. These unique mandibles are for specialized feeding. Skimmers cruise along with their lower bills slicing through the water. When they detect a prey item, they quickly snap their bill shut. Because of this tactile feeding, skimmers are quite capable of foraging in low-light conditions. They are active throughout the day but especially so during the crepuscular hours of dawn and dusk.

SPOTTED CATS OF THE SOUTH

A twenty- to thirty-pound spotted cats with a ringed tail, the ocelot is found from the tropical rain forests of South and Central America to the dry desert scrublands of South Texas. The species used to reach Arkansas and Louisiana, but now the remaining few dozen cats in the United States are all found in isolated patches along the Rio Grande Valley, especially near Laguna Atascosa National Wildlife Refuge, near Brownsville, Texas. In the northern part of their range, ocelots tend to deliver kittens in the fall. Breeding is less seasonal in other parts of the world. Highly nocturnal, the species spends the days in dense thickets of brush or perched in trees. The predators aren't too particular about their prey, feeding on rodents, birds, reptiles, and whatever else they can catch. Ocelot sightings are occasionally reported in southern Arizona, as is a much larger spotted cat—the jaguar.

HILLBILLY MANGO

Widespread but never abundant, pawpaws grow from southern New England and the Great Lakes to the Florida panhandle. Local names for the tree include hillbilly mango, Quaker delight, and Ozark banana. Yellow-green pawpaw fruits are visible in fall, but you'll have to beat the critters to them if you want to taste the sweet pulp. Pawpaw has an essence of the tropics, and the flora flavor, which has a hint of a yeasty aftertaste, is a niche market in the microbrewery industry. But don't expect to find pawpaws in your grocery produce aisle—the fruit bruises easily and has a short shelf life. Underripe fruits don't mature well off the tree either, so prime pawpaw season is short.

The tree also has a fan in the zebra swallowtail butterfly. Pawpaw is the host plant for these caterpillars, one of the few species to dine on the leaves. Most species avoid them due to high levels of acetogens. The caterpillars accumulate this toxin and, in turn, are less palatable to predators.

NEIGHBORLY TREES

We often think of trees as individuals competing for resources—every trunk for itself. But there is evidence that some trees are in it together or are at least responding to cues from their neighbors. Research shows that individual trees can receive signals from nearby plants and that this communication can transcend species. Associated with the host plant's root systems, mycorrhizal fungi are the conduits for much of this sharing and also benefit from the association. Even nutrients like carbon,

Tree Communication

nitrogen, and phosphorous can flow along this network. Some species have been shown to release chemical signals when they experience herbivory or pathogens, and in some cases, plants in close proximity put up chemical defenses of their own. In this way, larger hub trees may give smaller young trees a helping hand.

Not all communications are positive, though. Allelopathy signals can be harmful to the adjacent plant life. Numerous species, including pines, maples, and walnuts, release chemicals from roots, leaves, or buds that inhibit the germination and growth of understory cover.

SEPTEMBER 19

UNDERGROUND WATER

An aquifer is a layer of permeable water-bearing rock found underground. Aquifers that refill with surface precipitation are called unconfined aquifers, while confined aquifers are sandwiched between two layers of less permeable rocks. These groundwater-holding areas are like subterranean watersheds. The core of North America's largest aquifer, the Ogallala, takes up most of Nebraska. The

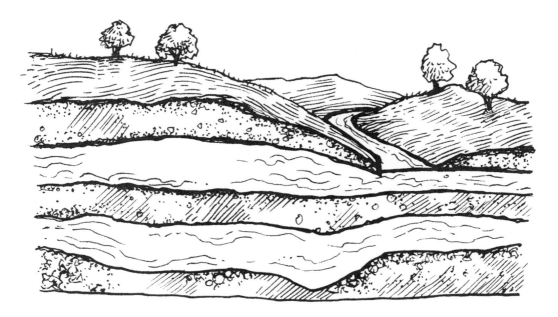

Aquifer

feature's boundaries spill over to seven adjoining states and stretch south to West Texas. Water has been accumulating in the Ogallala Aquifer for thousands of years.

In the last century or so, the aquifer has been heavily tapped for irrigation and drinking water. The aquifer's use currently exceeds its recharging rate, and according to the United States Geological Survey and other organizations, future trends look to be even worse for the water supply. The Ogallala Aquifer Initiative, a multiagency collaboration, is tasked with improving water quality and reducing aquifer water use in the region.

MOUNTAIN WATCH

While the National Aeronautics and Space Administration (NASA) needs citizen scientists to look over their photo albums (see July 3), the Appalachian Mountain Club needs volunteers to become the photographers. The organization's Mountain Watch and AT Seasons programs encourage hikers to upload the images they take and data they collect from the Appalachian Trail corridor to the iNaturalist app (see October 28). This geolocation-marked data helps map out the phenological changes along the length of the 2,200-mile trail that runs from Georgia to Maine. Photos from high points along the route, specifically at Madison, Lakes, Galehead, and Greenleaf Huts, are also used to document air quality and visibility information as a part of the Mountain Watch View Guides program. For thru-hikers and weekend hikers alike, being a citizen scientist is as easy as sharing a photograph.

FALL

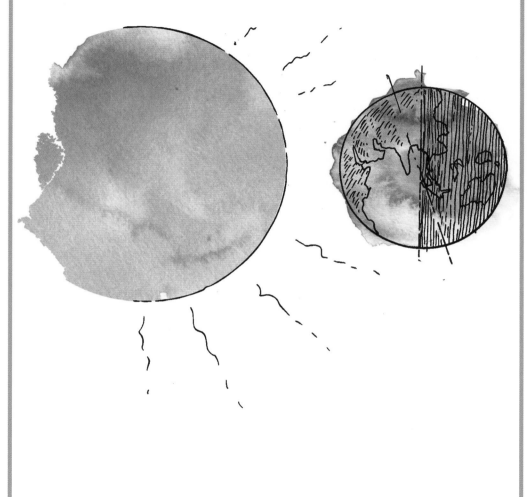

Autumn Equinox

AUTUMN EQUINOX

Above the equator, the Autumn Equinox marks the transition of the seasons away from summer. As winter approaches, Earth's tilt gradually points the Northern Hemisphere away from the sun during this portion of the annual orbit. The exact date of the equinox varies by a day or two because Earth's orbit around the sun isn't exactly 365 days long. Observed every four years, leap year helps correct this imbalance and keeps our seasons lined up with our calendars (see February 29). The full moon most closely associated with the Autumn Equinox is termed the harvest moon. Its added brightness aids farmers working extended hours to put up the fall crops. The Autumn Equinox brings a sense of urgency to plants and animals, too, as they prepare for winter.

(REIN)DEER OF THE NORTH

Caribou and their domesticated relatives, reindeer, are especially suited for life in the Far North. These members of the deer family have a short stocky build that, coupled with a double coat of hair, helps them retain body heat via countercurrent heat exchange, which offsets Arctic temperatures. During winter, their oversized hooves act as snowshoes, and their fleshy footpads shrink as stiff hairs grow in along the hardened hooves, helping to protect the animals from heat loss to the snow. Caribou roam far from season to season.

Barren-ground caribou, including the well-known Porcupine herd, make the longest migrations, traveling annually between calving grounds at the Beaufort Sea in the Arctic National Wildlife Refuge and their wintering range some 1,500 miles away. The timing of this annual migration is increasingly offset from the phenology (see January 1) of forage along the route because as the growing season advances, the caribou miss its peak and the herd produces fewer calves.

CARRY-OUT CARRION

The turkey vulture is an underappreciated bird, but a number of festivals around the country aim to change that. Fall is when these birds are celebrated in the Kern River Valley of California; Makanda, Illinois; and Athens, Georgia. In the springtime, the arrival of the first turkey vulture marks a transition away from winter for northern climes, and celebrations occur in Hinckley, Ohio, and Portland, Oregon. There is also a summer event at Coralville Lake in Iowa.

Perhaps vultures have gotten a bad rap due to an association with death. Sure, they feed on carcasses, but they are well suited for the task. Vultures can sniff out odors over a mile away. Their bald heads (red on adults, black on young turkey vultures) help to keep any rotting flesh from accumulating in feathers when they plunge their heads into bloated carcasses. With diets heavy in carrion, vultures have specific bacterial associations and powerful acids in their guts to aid digestion and protect the birds from foods that could be toxic to other critters. Nature's avian cleanup crews, vultures and condors should be celebrated every day. Give them a little tip of the hat next time you see one soaring overhead with outstretched, tipped-up wings.

❯ SEPTEMBER 24 ❮
CHICKEN OF THE WOODS

There are a couple of different mushrooms with poultry names out there, and many of them perch on the sides of trees. The *Laetiporus* shelf mushrooms, commonly referred to as chicken of the woods, are distributed nearly worldwide. Another common name, sulfur shelf, is equally as descriptive. These bright-yellow-orange fruiting bodies are bracket fungi that form platforms on tree bark. Characteristically polypore, the dense mushrooms have tubelike pores as opposed to gills on the underside.

> **SEPTEMBER 24, 1906**
> Following the creation of the Antiquities Act earlier in the year, Devils Tower in Wyoming becomes the nation's first national monument.

Some folks may experience minor reactions to eating them, but young chicken of the woods are generally considered edible, although they lack the rich flavor of many other species. (With any edible mushroom, it's always a good idea to try only a small bite the first time you eat it—reactions can vary, even with varieties that are widely edible.) Older shelf mushrooms tend to be brittle and less desirable for consumption. The hen of the woods is a related, mottled-brown shelf mushroom used in traditional Chinese medicine.

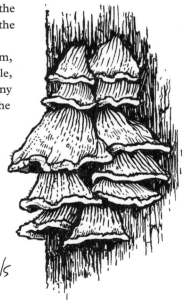

Chicken of the Woods

WILD RICE

Minnesota's state grain is wild rice. This grass species was first commercially cultivated in the 1950s but has a long history of being harvested from shallow lakes and slow-moving streams. Several Native American cultures consider the plant to be sacred. One traditional collection occurs from boats and involves tapping the grain seeds free with wooden sticks or poles called flails or knockers. The plants ripen gradually from top to bottom, so people visit the same patch multiple times to allow continued harvesting. After collection, the rice is dried and hulled. The grain has a delicious, almost nutty flavor, and is locally available in the fall in areas with a harvest. Wild rice is an annual plant, so seeds that sink to the bottom of the water will sprout the following year, ensuring a sustainable harvest. Texas wild rice, a protected species, is found only in a short stretch of the San Marcos River.

SWIMMING FOSSILS

Sturgeon are an ancient-looking fish, and rightly so. This lineage of cartilaginous fishes traces back a couple hundred million years to the Triassic Period. Instead of scales, the smooth-skinned swimmers have rows of bone-like plates, called scutes, along their elongated bodies. Mostly bottom feeders, they sport sensory barbel whiskers that aid feeding in dark depths. Their diet is heavy in benthic life, such as crustaceans, mollusks, insects, worms, algae, fish, or fish eggs. Some species, like Atlantic, green, and lake sturgeon, move between the oceans or the Great Lakes and large tributary rivers. Others, including pallid and shovelnose, make seasonal migrations within large freshwater river systems. Sturgeon are long-lived (up to one hundred years) and slow to reach sexual maturity. These factors, coupled with habitat loss and degradation from channelization and damming, have led to declines in all species in North America. Numerous conservation efforts are in place to bolster their populations. Hatchery-reared young fish are released into the wild in fall. Sturgeon festivals take place in Tennessee, Ohio, Michigan, Wisconsin, and Washington state. In parts of the world, especially the Caspian Sea, sturgeon are the source for prized

Sturgeon

caviar. The practice is more limited in North America, although in a few instances, sturgeon and related paddlefish eggs are collected for caviar in the Mississippi, Missouri, and Ohio River drainages. Limited spearfishing for sturgeon through the ice seasons is permitted in late winter in some northern regions.

SEPTEMBER 27
TURNING THE TIDES

The rise and retreat of the oceans is controlled by a number of factors, including the gravitational pull of the moon and the sun, coupled with the rotation of the earth. The largest tidal fluctuations are called spring tides, but they can happen during any season. Spring tides occur during new and full moon phases when the earth, sun, and moon are all lined up. The highest tides occur when the moon is between the sun and the earth at new moon and the gravitational pull of both the sun and the moon are combined. The largest tides in the world occur in eastern Canada at the Bay of Fundy and Ungava Bay. Here, tides reach fifty-five feet. Neap tides occur when the pull of the sun and the moon counteract one another, which happens seven days following spring tides.

END OF SEPTEMBER

Sea Otter Awareness Week, in the last week of September, recognizes the importance of these aquatic weasels in nearshore ecosystems. Various festivals, events, and activities throughout the week highlight conservation efforts and natural history of the species.

Though the Great Lakes do experience tides, the difference between highs and lows are just a couple of inches. Instead, barometric pressure, local precipitation, and even winds have a larger impact.

SEPTEMBER 28
FLY AWAY HOME

Whether you call them ladybugs or ladybirds, these spotted beetles are favorites—and not just with children. Six states have declared them their official insects—New York, Ohio, Massachusetts, New Hampshire, Tennessee, and North Dakota—while Delaware opted to proclaim them the state bug (because the First State has also recognized other insects: The monarch is the state butterfly and the stonefly is the state macroinvertebrate). Gardeners love them for their aphid-eating habits. The most abundant species, seven-spotted and Asian, are both nonnative, and many of the five native versions are declining.

Males and females look alike, and their bold colors—mostly reds, yellows, and oranges—and contrasting black dots or stripes serve as a warning for would-be predators. Ladybirds exude an irritating and smelly substance from their knee joints when threatened, though this reflex bleeding isn't enough to protect them from the assassin beetles, stink bugs, spiders, and toads that sometimes make meals of them. In the fall and early winter, ladybirds move to warm shelters for protection, including inside buildings.

QUAKING GOLD

Rocky Mountain National Park turns gold in the fall with leaves of aspens, *Populus tremuloides*, shimmering in the breeze. The quaking comes from flat-stemmed petioles (where the leaf attaches to the branch) that are at right angles to the leaves. Unlike most tree species, where each trunk is an individual, aspen clones are made up of multiple trunks sprouting from an extensive root system; groves can be made up of a single individual or of multiple plants. A single stand of aspen in Utah's Fishlake National Forest stretches over one hundred acres and weighs an estimated 14 million pounds; it is the most massive and one of the oldest organisms in the world. Root systems can easily be from five to ten thousand years old in clones, but the age of this Pando grove has been dated to eighty thousand years old.

The white bark of aspen shows a tinge of green chlorophyll that allows the trees to photosynthesize throughout the winter. These trees are important habitat and forage for moose, elk, black bear, snowshoe hare, ruffed grouse, beaver, and a variety of songbirds. Bigtooth aspen, *Populus grandidentata*, is the species on drier sites in the East.

MAST CROPS

Oak, hickory, beech, and other trees produce nuts in cycles. Some years hardly any of these seeds are grown. Other years there are bumper crops of this mast—the edible seeds of woody plants. The cycles of masting aren't entirely understood. Weather may play a role as stands of trees tend to show synchronized masting. Tree physiology is also involved. The production of seeds is essential, but it is also energetically costly; in heavy mast years, trees grow less. The predator-prey dynamics also play out. These mast cycles are exploited by wildlife, which benefits the plants.

Just as herding protects individual animals, an abundant supply of seeds means some seedlings are sure to survive despite the heavy consumption. Animals also help the plants by dispersing the seeds. Populations of white-footed mice have been shown to increase following mast years and decline between the peaks in production, sharing a cyclical pattern similar to the dynamics of the predator-prey cycles.

Aspens

AN IMPRESSIVE FLIGHT

In contrast to their dapper black-and-white breeding plumage, blackpoll warblers are more generic looking in the fall. Orange legs are perhaps the best field mark to look for during any season. Even if you can't identify these little fliers, they deserve your respect. In the spring, they move from South America, over the Caribbean Islands, and throughout eastern North America before fanning out across boreal Canada and Alaska.

Blackpoll Warbler

Their fall journey is even more epic: Blackpolls move east to the Mid-Atlantic before staging an impressive transoceanic flight. The half-ounce birds travel nonstop for up to three days and 1,800 miles. Before takeoff, the warblers put on extra weight in the form of fat reserves, functionally topping off their gas tanks. They also take advantage of prevailing southern winds, flying on the leading edge of cold fronts. Blackpoll warblers from the western breeding populations have longer wings than their eastern counterparts, a trait that may be associated with the added miles they put on each year.

A LOT OF GALL

Although they are usually harmless to the plant, swollen stems or blistered leaves can both be signs of insect damage. These abnormal growths, called galls, are often caused by larval stages of invertebrates, including gall wasps, gall midges, aphids, or plant mites. These growths can be easiest to spot this time of year. Goldenrod is the host plant for the widespread goldenrod gall fly. Gall flies deposit their eggs directly on goldenrod plants, where they hatch. Hatched larvae induce gall formation. The larvae overwinter inside the galls before undergoing metamorphosis into adults the following spring. The firm gall bulbs offer some protection from the elements and predation, but the larvae aren't entirely immune from them either. Though the invertebrates can tolerate freezing temperatures, not all individuals will survive.

OCTOBER

The entire month of October is bat appreciation month for Bat Conservation International. Festivities culminate during the final days of the month with Bat Week ending on Halloween.

Black-capped chickadees and downy woodpeckers seek out galls for foraging, pecking the growths open to reach the fly larvae. Some species of parasitic wasps target these goldenrod gall flies too.

NINE PLANETS MINUS ONE

For folks of certain generations, the solar system once had nine planets. In 2006, space lost a planet. Well, *lost* may not be the right word. Pluto is still there, approximately 3.67 billion miles from the sun, but it has been reclassified as a dwarf planet. Though this change was emotionally difficult for many, the science is solid. Pluto orbits the sun and is massive enough to be rounded by its own gravity, so it meets two of the three requirements to be recognized as a planet by the International Astronomical Union. However, it doesn't pass the final planet hurdle. It doesn't clear debris from its orbital neighborhood; instead other objects, including asteroids, comets, and other celestial bodies, are still nearby.

Pluto is a relatively small chunk of ice and rock in the Kuiper Belt. The nearest planetary neighbors to the Kuiper Belt, Neptune and Uranus, are ice giants. Moving closer toward the sun are the gas giants Saturn and Jupiter. The remaining four, Mars, Earth, Venus, and Mercury, are rocky planets. The planets lined up end to end would fit nicely between Earth and our moon. Sure, the reclassification of Pluto is a bummer, but can you imagine how bad Eris and Ceres feel? They are also dwarf planets, but you've never even heard of them, have you? At least Pluto still has some notoriety and name recognition.

PUTTING ALL THE NUTS IN ONE BASKET

Acorn woodpeckers have an especially complex social system, with multiple breeders in each colony. These birds are generally related, and multiple females lay eggs in a single nest. Older siblings often remain within the group's territory for years and will help raise young and collect food. A lot of animals stockpile nutrients by caching foods, but acorn woodpeckers store all of their nuts in a single wooden basket. These granary trees (or granary power poles, fence posts, wooden walls, etc.) are a group effort. Each fall the birds all pitch in to stock the cupboards, mostly

OCTOBER 4, 1946

Gifford Pinchot passes away at the age of eighty-one. The first head of the Forest Service, he helped shape Teddy Roosevelt's thinking on sustainable timber harvest. Pinchot also served as governor of Pennsylvania and founded the Yale University School of Forestry. A national forest in southwest Washington state is named for him.

Acorn Woodpecker

with acorns, but they also store other nuts. A single granary can have fifty thousand seeds tightly wedged into divots carved out by the woodpeckers. As the nuts dry and shrink, the birds move them to tighter holes.

In addition to the stashed supply, acorn woodpeckers have a well-rounded diet throughout the year, including insects, fruit, sap, and even flower nectar. Look for these red-capped birds with comically bright eyes and boldly white faces in oak and mixed woodlands along the Pacific coast states and the Southwest east to Texas.

▶ OCTOBER 5 ◀
HOARDERS

Food hoarding is common in nature; many species have a single cached stash. Think of a red squirrel, chattering defiantly. It's likely defending a food supply and agitated that you're approaching too close. This food-caching behavior is called larder hoarding. You may find a midden of pine cone scales piled up surrounding the squirrel's dinner table. Even the big cats will hoard a meal for a week or two. Mountain lions consume roughly twenty pounds of meat at a time, so they bury deer carcasses under piles of dirt and leaves between feasts.

Other species, the scatter hoarders, spread their food goods far and wide, a behavior found in both birds and mammals. Jays collect seeds and nuts, one at a time, from feeders or the wilds. Most squirrels do the same thing. They bury a nut here, another over there and return later to eat these stored goods. Thousands of morsels are tucked away seemingly randomly across the landscape, but this behavior is intentional. In the fall, gray squirrels often consume white oak acorns while caching red oak nuts. White oaks sprout in the fall, so they'd lose nutritional value for the squirrels over winter. Red oak acorns don't sprout until the spring, making them the better choice for storage. Scatter hoarders also affect the ecosystem in a meaningful way; hoarded seeds that they overlook may sprout into new plants.

▶ OCTOBER 6 ◀
SHELL MIDDEN

Accumulations of discarded shells can be more than just heaps of archaeological kitchen scraps. In many locations, shells were used deliberately to construct mounds or rings, often for ceremonial functions. Expansive groupings of midden-mound complexes are collectively termed shell works. For example, the Calusa chiefdom in the Ten Thousand Islands region of southwest Florida was a collection of many small villages with Mound Key serving as the capital of the Calusa kingdom; the most extensive mounds here stretch nearly the length of a football field and tower thirty feet or more.

Some of the mounds in the area are replicated and arranged in a layout suggestive of the settlements' shared community structure. Canals were constructed for navigation and to potentially serve as ponds for fish and shellfish or to store water. The Calusa tribe utilized the resources of the area by fishing and foraging, and they used shells as hammers, picks, scrapers, jewelry, and ornaments.

OCTOBER 7

A NATIVE MONSTER

The largest lizard native to North America, the Gila monster tips the scales at a hefty four pounds and stretches to nearly two feet in length. They are found only in the southwestern states, mostly Arizona, and northern Mexico. In 2019 they were declared the state reptile of Utah. These lizards spend much of their lives underground and are most active during the spring and again during the late-summer rainy season. They feed just once or twice a year, usually eating eggs or small mammals and storing fat in their tail to get them through lean times.

These black tiger-striped beauties can appear orange, yellow, or even pinkish, but be sure to observe them from a distance and resist the temptation to pick one up. Although the dangers of Gila monsters have been widely exaggerated (they don't spit toxins, nor can they jump up and bite you in the face), they are venomous. And while a bite is rarely fatal to humans, the painful experience is best avoided for both the biter's and the bitee's sake. Gila monsters tend to bite hard and are reluctant to let go, instead continuing to gnaw and gnaw. Interestingly, a synthetic protein based on the saliva of Gila monsters is being tested as a treatment for type 2 diabetes.

OCTOBER 8

ENDEMIC

Diseases are said to be endemic, or restricted, to a region, but the term also applies to plants and animals: endemic species are those found in certain geographic locales and nowhere else. Scale is important when discussing endemics, be it a watershed, a state, or a nation. Isolation is a major factor that leads to a species becoming endemic. Islands, where life is separate from mainland populations, host many endemics. Hawai'i, for example, is home to numerous endemics, including almost fifty living species or subspecies of localized birds. Similar to islands, mountain ranges are surrounded by habitat barriers that isolate plants and animals, and each range can have unique life forms.

Aquatic specialists show this speciation too. The pupfish of the Southwest all share a common ancestor from when prehistoric lakes were interconnected. Separation has now led to at least thirteen endemic species found in scattered locations. This isolation can make endemics especially vulnerable to extinction.

Gila Monster

PREGNANT FATHERS

Seahorses perform elaborate courtship cere-
monies that can stretch for several days. It is
thought that these displays help synchronize
the reproductive states of males and females and
reinforce the pair bond. There are four phases of
seahorse foreplay. Stage one is initiated a couple of days before copulation and involves reciprocal quivering. The pair spends the night
apart but come together after dawn for a few minutes of quivering.
As courtship nears, the females assume a pointing position. Males
respond by pumping and quivering at first, and eventually they
also posture with the pointing display. Females use oviposi-
tors to lay eggs directly into the brood pouches of males.
Gestation lasts a few weeks on average, and then miniature seahorses are released from the pouches of
the males. However, seahorses aren't going to win
any father-of-the-year awards. The tiny young are
immediately on their own, receiving no additional
parental care. Like the adults, the slow-moving
seahorse babies rely on camouflage to keep them
safe from predators.

    If you happen to find a free-range seahorse, you
can report it to Project Seahorse's iSeahorse program,
a citizen science initiative to document the species across
the planet.

Seahorse

GINSENG HUNTING

Poaching is a growing concern for a number of plant species, including ginseng. Used for centuries,
the plant was one of North America's first exports. Said to be an anodyne for wide-ranging ailments,
the American plant and its Asian counterpart both fetch top dollar. Commercial harvest is permitted
and regulated in many states and provinces, although many public lands are off limits to collecting.

Ginseng is found in the Midwestern and eastern states and southern Canada. The lore of ginseng is especially rich in the Appalachian and Ozark Mountain regions.

Mature plants sport red berries, but collectors are after the roots, which they harvest in the fall. Many states require collectors to plant ginseng seeds to replace the harvested plants.

OCTOBER 11
RIVER OF RAPTOR MIGRATION

Riding favorable wind currents, raptors move south in pulses. The timing is different for each species, so keep your eyes to the skies throughout the fall. Broad-winged hawks often move through an area in a week. Other species string out migration movements throughout fall. Hawkwatch International sets up hawkwatch sites across the continent, often on mountain ridges or coastal corridors, where numbers are tallied each day. These sites can be great places to visit to experience the river of raptors firsthand. From late September to late October, the viewing platform at Veracruz, Mexico, can tally up to one hundred thousand raptors and vultures in a single day. But not all raptors head that far south. Rough-legged hawks, for example, are winter visitors across much of the US.

Around 45 percent of raptor populations migrate. Some groups, including Swainson's hawk, shift entirely from season to season. Other species, like the red-tailed hawk, have both resident and migratory populations.

OCTOBER 12
THE RUT

The mating season for the ungulates (hooved mammals) is called the rut. Each fall, rut rituals play out as males and females come together. Timing varies by location, but for some species, like bison, the rut can start as early as July. It is usually going strong for pronghorn and mule deer by September. For others, like white-tailed deer and elk, breeding continues until November or later. Bugling elk are one of the most symbolic signs of the rut. Males will seek out females, sometimes by gathering large harems of cows or does, and displaying males try to intimidate competitors and to impress the females.

Occasionally the posturing turns physical. Sheep rams head-butt. Deer and elk tussle, with antlers intertwined. Rarely are these encounters lethal. The struggles of the rut do wear on the males, however, and many enter winter with their bodies in poor condition as a result of their efforts.

FOSSIL FISH

The high sagebrush steppe of southwestern Wyoming looked far different during the Eocene Epoch some 40 million years ago. At that time, the region was covered by ancient subtropical lakes, including Fossil Lake, Lake Gosiute, and Lake Uinta. Collectively known as the Green River Lakes system, the region was under near-perfect conditions for the fossilization of fish. The cool, deep, and calm waters and fine-grained lake sediment led to the preservation of numerous fully

OCTOBER 13, 2010

National Fossil Day is first celebrated, during Earth Science Week. The day draws attention to the scientific and educational values of fossils and is now celebrated by nearly three hundred organizations and agencies.

articulated skeletons. The geologic layer known as the Wasatch Formation preserved many Eocene species, which help reveal the story of the terrestrial life around the Green River Lakes System.

Fossil Butte National Monument showcases the region's world-class paleontological heritage. To date, the number of fossil species identified in the area includes twenty-seven fish, ten mammals, thirty birds, fifteen reptiles, two amphibians, a number of invertebrates, and several plant specimens. Fossil Butte isn't just living in the past. Around 150 living animal species roam the monument landscape these days, although it no longer harbors any fish.

"DING" DARLING DAYS

Jay Norwood "Ding" Darling (1876–1962) was an American cartoonist whose drawings often depicted conservation and politics. He gained a following in the early twentieth century as the cartoonist for the *Sioux City Journal*, in Iowa, and later for the *Des Moines Register*, where he signed his work "Ding," using the first letter of his last name with the last three letters. His cartoons made such a powerful impact that in July 1934, President Franklin D. Roosevelt appointed him the director of the Bureau of Biological Survey (now the US Fish and Wildlife Service).

During his eighteen-month stint in that role, he launched the Federal Duck Stamp program (see November 16) and designed the first duck stamp himself. A duck stamp is required for most waterfowl hunters, and the funds generated from the sale of the stamps are used for wildlife and wetlands protection.

Every year, artists around the country compete to have their wildlife artwork on the stamp. You can usually meet the winners of the Federal Duck Stamp Contest during "Ding" Darling Day, held annually in mid-October at J. N. "Ding" Darling National Wildlife Refuge on Sanibel Island, Florida.

HEDGE BALLS

Related to mulberries and figs, Osage orange is native to the Red River Valley along the Texas-Oklahoma border. Widely cultivated, the thorny tree was planted extensively as a hedgerow species, so it is now well established in pockets throughout the country. In the fall it produces unique fruiting bodies, softball-sized hedge balls that are greenish yellow and bumpy in texture. Hedge apples, as they are sometimes called, are sold at farmers markets as pest remedies. Many folks believe placing them around the foundation of their homes or in basements will keep insect numbers down. Research has shown that compounds in the plant can exhibit some insecticide qualities, but the concentrations found naturally in this fruit don't.

What is valuable from the tree is the wood. Hard and durable, Osage orange has been used to make fence posts, furniture, tool handles, and prized archery bows. One regional name for it is *bois d'arc*, French for "bow wood." The yellow-orange wood fades over time, but the plant can also be used to make a vibrant dye.

OCTOBER 15, 1966

On Michigan's Upper Peninsula, Pictured Rocks National Lakeshore is established along the southern shores of Lake Superior.

Osage Orange

GOING UP

Migration can take many forms. Altitudinal migration is when critters head down the mountain for winter, but a reverse altitudinal migration is a bit perplexing. For the most part, big game herds winter in the foothills and lowlands in the mountains, but a few individuals head up the mountain when weather hits. The windswept ridges can remain snow-free, exposing vegetation. Birds also shift on the mountain landscape. Plenty of songbirds are wintertime visitors to feeders but then move back upslope for the breeding season.

The grouse formerly known as blue, now separated into two species (sooty, which is near the coast, and dusky, which is inland), bucks the migration trend. After spending fall in the deciduous forests feeding on buds and catkins, the birds move up to higher elevations for winter. These birds feed on tree needles and roost in conifer stands until spring rolls around.

BIRD BRACELETS

None other than John James Audubon placed the first leg band on a bird in North America. The eastern phoebe was marked with a thin thread harmlessly tied around the leg. The bird migrated south for winter but returned the following spring. Bird banding is now regulated by a division of the United States Geological Survey. The Bird Banding Lab issues permits and maintains the database of all birds tagged. Aluminum leg bands are stamped with unique codes, much like a human social security number. Banders collect information from the birds, including species, body measurements, and plumage characteristics. The age and sex are recorded when determinable. Then the birds are released.

Organizations can often use volunteers for bird banding efforts, but that's not the only way to be involved. If you encounter a banded bird, report the number online (www.reportband.gov). The numbers are tiny but can sometimes be deciphered from photographs, particularly of birds visiting feeders. The Bird Banding Lab will let you know where the banded bird came from, and they'll inform the researchers where the bird ended up.

FOX IN A TREE

Gray foxes are sometimes mistaken for red foxes. Both species can appear mottled gray with red, especially along the neck, sides, and legs, but you can tell them apart by their tails: Look for a dark stripe and tip on the gray fox's tail. Red foxes have white-tipped tails. The gray fox is a canid with a somewhat catlike feature—they have semiretractable hooked claws.

OCTOBER 18, 1972
The Clean Water Act is enacted to provide the basic structure for regulating pollution in the waters of the United States.

Found from coast to coast with the exception of the northern Rockies and western Great Plains, this small fox is widespread but rarely encountered due in part to its nocturnal habits. The species spends a considerable amount of time in underground burrows, caves, crevices, and hollow logs. Gray foxes will also climb trees on occasion to rest, eat, or avoid predators. Gray foxes are omnivores and eat everything from insects to fruit. In the fall, young foxes disperse and move away from their parents. The Channel Islands are home to an endemic species of gray fox (see October 8). The island fox is even smaller than the mainland type and is more crepuscular in nature.

Gray Fox

LONGEST FLIGHTS

The Arctic tern could just as easily be called the Antarctic tern. Clocking up to fifty thousand miles per year, this bird species has the record for the longest annual migration. Arctic terns breed in the Far North across North America, Europe, and Asia. These breeding colonies are the best bet for finding one; otherwise they are far out to sea. After the chicks fledge, the birds move south—extremely far south. They spend the northern winter along the edge of the Antarctic ice pack for the Southern Hemisphere's summer. These birds probably experience more daylight than any other over the course of the year. Arctic terns live on average thirty years, so it's not unreasonable to figure the birds fly the equivalent of three roundtrip journeys between the earth and the moon in their lifetime.

OUT TO SEA

Pelagic refers to the open ocean as well as the birds that spend the majority of their lives far out at sea. Going birding by boat is the best way to see these species, and potential seasickness is a small price to pay if you want to see amazing birds. Pelagic birding trips are typically full-day excursions, although overnight multiday trips are also common. Tube-nosed birds, including albatross, shearwaters, petrels, and storm petrels, are common targets. Albatross have the longest wingspans of any birds, but all of the pelagic species are adept fliers. Storm petrels sometimes appear to be dancing on the surface of the ocean as they feed.

The tubed noses of this order of birds are adaptations that make living in salt water possible. They excrete excess salt through their nasal glands. Unlike many birds, this group has the ability to smell, which helps them find food and locate their nests in expansive colonies. Similar open-water boat tours on the Great Lakes are growing in popularity, although you aren't going to see an albatross on such an outing. These tours are often looking for waterfowl, jaegers, and rare gulls.

MOUNTAIN BOOMERS

The origin of the name mountain boomer has been lost to history, but this creative term refers to the state reptile of Oklahoma, the collared lizard. Common and widespread in dry, open terrain from Missouri to Southern California, the species is variable in color, with the males often displaying a vibrant blue-green with yellow markings and a lighter head offset by distinctive collar stripes. Females have subtler coloration. The species can grow to over a foot long, and a lot of that is tail. These lizards can't shed this appendage like some other species. The tail is essential for balance, since collared lizards can stand up and run on hind feet for short distances. They've been clocked at sixteen miles per hour. Collared lizards will soon be hunkering down for the winter season in the northern regions of their range but could remain active on the warmest days throughout the year in the South.

▶ OCTOBER 22 ◀

WE ARE THE CHAMPION . . . TREES

Known as the redwood park of the east, Congaree National Park in central South Carolina protects the largest intact stands of bottomland hardwood forests and many of the biggest individual trees. The Congaree and Wateree Rivers send nutrients throughout the floodplain, supporting such giants. While there are no actual redwoods at Congaree, there are some mighty contenders. The park is home to the tallest loblolly pine in the world, at 167 feet. Other gold medal trees include a 157-foot sweet gum, a 154-foot cherrybark oak, a 135-foot American elm, a 133-foot swamp chestnut oak, a 131-foot overcup oak, a 127-foot persimmon, and a 125-foot laurel oak.

The forests of Congaree support impressive diversity. Twenty-three types of vines have been identified here, along with 170 species of birds, nearly 50 types of fish, and around 30 species each of mammals, reptiles, and amphibians. The most famous of the park's wildlife inhabitants, though, is an insect, the synchronous firefly (see May 29).

Loblolly Pine

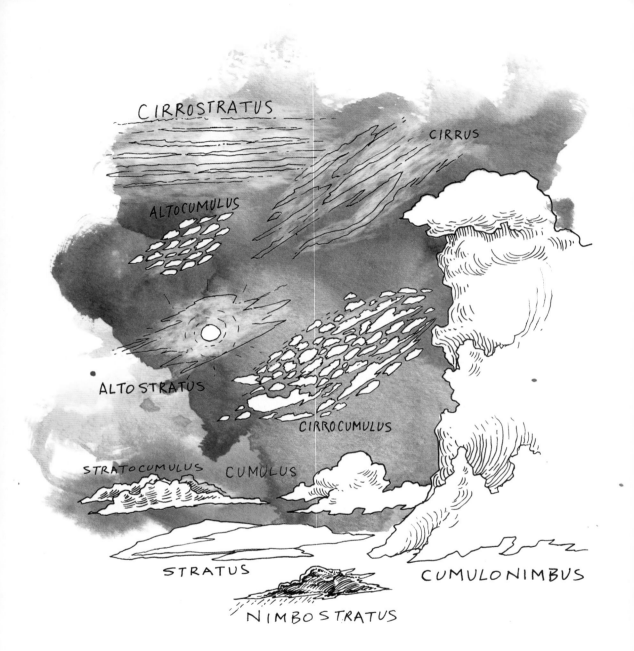

CIRROSTRATUS

CIRRUS

ALTOCUMULUS

CUMULONIMBUS

ALTOSTRATUS

CIRROCUMULUS

STRATOCUMULUS CUMULUS

STRATUS

NIMBOSTRATUS

Cloud Forms

HORSETAIL

Horsetail plants are not grown from seeds; instead this ancient group of primitive vascular plants reproduces via spores. The plant looks similar to bamboo with its distinct jointed segments, and it's hard to resist pulling a few lengths apart whenever you find some. Around 350 million years ago, during the Devonian Era, the *Equisetum* horsetails would tower to the same heights as the redwood trees of today. Modern *Equisetum* rarely top out above ten feet tall; most are three to five feet in height. The majority of the twenty or so remaining species are found in wet areas. The stems are high in silica, and the silicates are thought to provide structural support for the hollow stems. Sometimes called scouring rush, horsetail is often used in landscapes and pondscapes, although take care if you choose to plant it as it can prove invasive and take over an area.

WATCHING THE CLOUDS FLOAT BY

The mood of the sky is set by the clouds, as is the weather. As water droplets or ice crystals form around nuclei in the atmosphere, clouds are created. Cooler temperatures and changes in air pressure facilitate this process. Clouds are categorized by both their shape and by their elevation in the sky. Cumulus, stratus, and stratocumulus clouds are all below six thousand feet. Cumulus clouds are the ones that take you back to childhood by morphing into all kinds of imaginative shapes. Stratus clouds are fog-like and cast a dreary gray overcast. Stratocumulus clouds are characterized by lumpy contours.

The alto cloud types, including altocumulus and altostratus, are similar but form at mid-altitudes up to twenty thousand feet. Sometimes called mackerel sky, cirrocumulus cloud tufts appear scale-like across the sky, as opposed to cirrus clouds that are wispy streaks and referred to as mare's tails. Cirrostratus are also thin clouds and can cause a halo around the sun or the moon. They can be precursors to precipitation. The big billowing storm clouds are termed cumulonimbus, but they should be called "cumulo-menacing" clouds as they can cause extreme weather.

PSEUDO GILLS

There are a number of species of chanterelle mushrooms found in the Northern Hemisphere and south to Central America and Africa. These fungi can fruit from summer to late fall. Aptly named, winter chanterelles materialize in the woods later than many other mushrooms. This group of mushrooms is mycorrhizal, growing in association with tree roots, which makes locating a patch year after year more reliable, but also makes them susceptible to overharvest and exploitation. Despite being popular in culinary circles, the fungus doesn't do well under artificial growing conditions.

The funnel-shaped mushrooms can be orange, yellow, or nearly white. False, or pseudo, gills are one of the identifying characteristics, but there are inedible doppelgänger species to be aware of. An almost apricot aroma can sometimes be detected from chanterelles. As a bonus for foragers, with the exception of pig ears, chanterelles usually don't get too wormy, though they may come pre-nibbled because many wildlife species enjoy mushroom appetizers.

BUCKEYES

Buckeyes

Buckeyes aren't just for Ohio. Half a dozen species are native to North America. The most widespread, the so-called Ohio buckeye, *Aesculus glabra*, can be found from Pennsylvania to Alabama and west to Iowa. Despite sometimes being called horse chestnuts, all parts of the buckeye plants, including the well-known nuts, are high in tannic acid and inedible to livestock and people, although many species of wildlife appear unaffected by consuming them. Broken limbs and bruised seeds release an unpleasant odor, which has led to another common name: the fetid buckeye.

In the springtime, buckeyes have showy flowers. During the summer, the namesake nuts are enclosed in a leathery husk. When the husk is shed, the dark-brown orbs with tan spots are said to resemble the eyes of deer. (The combination of colors is also recreated in the sweet buckeye treat, peanut butter balls dipped in chocolate!) In the fall, the leaves blaze in scarlet reds, oranges, or yellows.

GARKANSAS

For a state known as the Natural State, Arkansas was a bit late to the game of declaring an official fish. The state finally embraced an underdog in 2019, thanks to some enthusiastic groundwork by Henry Foster. The ten-year-old petitioned the legislature to choose the alligator gar because it is tough and unique. He also felt like the gar would make Arkansas stand out from a crowded school of official trout, bass, walleye, and catfish. It was a gargantuan effort to bring recognition to what's long been heralded as a trash fish. Despite research to prove the contrary, the top predator gets a bad rap for harming fisheries. There was a bit of backlash, but the native alligator gar eventually prevailed. And it is a worthy choice. Besides Troutkansas and Basskansas just don't roll off the tongue like Garkansas does.

CATALOGING NATURE

Perhaps the easiest way to become a citizen scientist is to join iNaturalist (www.inaturalist .org) or eBird (www.ebird.org). These programs encourage people to record the species they find. It's as simple as that. And in the case of iNaturalist, you can snap a photo and ask other members help you identify anything you are unsure of. iNaturalist is a joint collaboration between the California Academy of Sciences and the National Geographic Society. The Cornell Lab of Ornithology administers eBird. Both programs also offer troves of data to delve into. When you're planning to travel to a new area, these sites can serve as virtual field guides. If you wonder about the expected arrival and departure dates for a bird in your county, eBird can plot the data for you. Curious about the mammals of Mongolia? iNaturalist is a good place to start your armchair exploring.

DEEP ROOTS

Tallgrass prairies once stretched from southern Canada south to Texas, and from Indiana west to Kansas. The ecosystem covered 170 million acres, but the vast majority of it has been plowed under and replaced with a growing urban footprint and fields of corn and soybeans. Native tallgrass prairies are home to an impressive diversity of five hundred species of plants, not just grass. A number of wildflowers bloom in waves throughout the growing season. With so much happening on the surface, it's easy to overlook that most of the biomass in a prairie is underground. The eight-foot-tall big bluestems and compass plants have root systems that grow down twice as deep into the soil as the stems grow up into the sky.

Historically, bison roamed the tallgrass landscape, and today, Tallgrass Prairie National Preserve (Kansas), Neal Smith National Wildlife Refuge (Iowa), and Midewin National Tallgrass Prairie (Illinois) are all home to herds of the large grazers. In the fall, these preserves also offer the rare opportunity to walk through tunnels of grass swaying at head height.

FOREST WEASEL

Fishers, medium-sized carnivores in the weasel family, are larger than the similar American marten and mink. Despite their name, they target small mammals like snowshoe hares, porcupines, and squirrels—not fish. They also dine on snakes, birds, mushrooms, and fruit when available.

Fishers were extirpated from much of their historic US range, in some cases a couple hundred years ago, by overharvesting. Their dense chocolate-brown fur was prized and brought top dollar for early trappers. Loss of forested habitat also took a toll on the species as woodlands were fragmented in the Appalachian Mountains, New England, and the upper Midwest. Reintroduction efforts have taken place in nine states, and populations seem to be rebounding. Some Wisconsin fishers were moved to Montana and Tennessee, and Canadian fishers now roam Washington's Olympic and Mount Rainer National Parks. West Virginia received fishers from New Hampshire, and this population has expanded, recolonizing historic habitats in Pennsylvania, Maryland, and Virginia.

> OCTOBER 31 <

HALLOWEEN SPIRITS

Your best bet to find what is known as a Halloween pennant dragonfly on Halloween is in Florida, but during the late-summer and fall seasons, the boldly marked odonates are found from New Mexico to Maine. Halloween pennants are orange or yellow, with distinctive black-striped wing markings. Like others in the pennant group, these dragonflies perch at the end of plants, fluttering on the breeze—like pennants. Dainty fliers, the pennants are sometimes mistaken for butterflies at first glance. They are keen insect predators, as both adults and as aquatic naiad larva. Halloween pennants are equally as likely to be found at water's edge and in fields. Dragonflies aren't the only species with a bit of Halloween spirit. Bright-orange jack-o'-lantern mushrooms grow on decaying stumps and roots of hardwood trees in the East. A similar-looking species in

OCTOBER 31, 1994

Already recognized as national monuments, Death Valley and Joshua Tree both become national parks on this date.

Halloween Pennant

Europe is known for having gills that glow in the dark. Claims of this bioluminescence in the North American species may be true, or they may be a figment of the imagination.

FLIPPING LAKES

Like layers of cake and frosting, water can be categorized in distinct zones: the littoral, limnetic, and profundal zones. The littoral zone is the ring of water near the shoreline, whether it be in a pond or lake. In freshwater ecosystems, emergent vegetation like cattails, rushes, and sedges, as well as submerged aquatic vegetation, grow in these shallows. This habitat band is critical to many species for feeding, protective cover, egg-laying, and as aquatic nurseries. The limnetic zone is the open water, away from the shoreline. This zone produces much phytoplankton and zooplankton. In deeper bodies of water, the profundal zone is below where the sunlight penetrates, extending down to the benthic zone along the bottom sediment.

These layers of water have varying temperatures and oxygen levels. They also have different densities, which is why lakes freeze over (see November 22). Surface water heats up in the summer, and deep water remains relatively cool. Eventually the temperature gradient changes as surface waters cool in the fall, which causes the waters of the lake to turn over with the deep waters rising to the surface. The process of lake waters flipping repeats in the spring.

THE TREES' KNEES

Bald cypress trees, the state tree of Louisiana, are one of the few deciduous conifers. They shed their needles annually. But they are also known for another mysterious feature. If you spend any time in a cypress swamp, you can't miss the knobby stumps sticking up out of the water. These cypress knees are a part of the trees' root systems. Their function has puzzled folks for generations. One theory is that the knees are pneumatophores, or aboveground roots that help gather oxygen for the submerged trees, but this idea seems unlikely.

NOVEMBER

Manatee Awareness Month is celebrated in November. The Save the Manatee Club was established in 1981 by singer-songwriter Jimmy Buffett and former US Senator Bob Graham, who was the Florida governor at the time. (For more on manatees, see December 29.)

Another idea is that the buttresses collect soil and reduce erosion, thus providing stability for the giant towering trees. Cypress trees growing in more upland sites tend to have fewer knees, evidence that supports the latter.

Corkscrew Swamp in southwest Florida is home to truly impressive cypress specimens. Other flooded timber sites along North Carolina's Black River and in pockets of Louisiana were spared from saw blades, and these areas still hold trees estimated to be hundreds of years old.

NORTHERN LIGHTS

Especially for the northern tier of the planet, winter nights that are clear, cold, and crisp can provide ideal conditions for catching a glimpse of the dancing lights known as Aurora Borealis. Northern lights are basically like neon signs of the night sky. They occur when flares of charged solar particles strike atoms within Earth's atmosphere. The electrons in the atoms release light as part of this reaction. Different atmospheric gases cause the different colors, all of which can appear as sheets, arcs, or spirals of light along Earth's magnetic field.

Many resources are available for forecasting northern lights, but it never hurts to keep an eye on the night sky. The northern lights are fairly common as far south as central Canada, but they can dip down into the northern tier of states, including Washington, Montana, the Dakotas, the upper Midwest, and New England, fairly regularly. The northern lights are an ecotourism boost of sorts. Many travelers seek out a chance to see the lights, and it is considered good luck to conceive a child under the astronomical phenomenon.

DABBLE OR DIVE

Ducks can be classified in two categories: dabblers and divers. They all have duckbills and webbed feet, but the similarities end there. Dietary preferences are revealed in the shapes of the birds. Dabblers, like mallards, teals, pintails, wigeons, and gadwalls, feed in the shallows by tipping over and sticking their duck butts in the air. They eat mainly submerged vegetation. The dabblers ride high in the water. Their legs are positioned near the center of their body, and they are more comfortable on land than on water.

Divers are efficient swimmers and more often feed on aquatic prey, including fish, amphibians, and mollusks. Some of the divers include canvasback, bufflehead, scaup, goldeneye, scoters, eiders, and mergansers. Divers float lower in the water and are rarely seen on land. Their legs are set farther back on their body, near the tail end, to aid in swimming—at the expense of land maneuverability.

Northern Lights

LOST BIRDS

The dodo, once found on the island of Mauritius
in the Indian Ocean, is the unfortunate poster child
for extinction. But what about North American species?
Several species of native birds are now found only in museums.
Todd McGrain of the Lost Bird Project created sculptures
of five such birds. Passenger pigeons were likely once
the most abundant bird on the continent, but market
hunting, paired with habitat loss, caused their demise.
The last passenger pigeon, named Martha, perished in the
Cincinnati Zoo in 1914. The Carolina parakeet is the only par-
rot endemic (see October 8) to eastern North America; the last wild
bird was killed in central Florida in 1904, and the last captive bird died in
1918. Curiously, it was held in the same aviary cage that housed the final
passenger pigeon. The heath hen, a prairie chicken relative, went extinct
in Massachusetts in 1932, while the great auk and Labrador duck were
declared extinct by the mid to late 1800s. The Lost Bird Project sculptures are on permanent display in the places where the species were
last documented in the wild.

Carolina Parakeet

WHALE MENOPAUSE

On average, female humans enter menopause between the ages of forty-five and fifty-five years. Well
before the end of their lives, hormonal changes lower their chances of reproducing. This physiological state is practically unheard of in the animal world. Primates, for example—including our closest
living relatives, chimpanzees—don't experience menopause. Only a handful of nonhuman mammals
have been identified as undergoing menopause, all of which are toothed whales, including orcas,
short-finned pilot whales, beluga whales, and narwhales.

For the whales, shared resources and linked genetics may play a role in these grandmother whales
that no longer reproduce. If the older whales were to continue reproducing, their offspring would
compete for resources with their genetic grandchildren. But there is still some debate on this theory
since elephants and other long-lived mammals in matriarchal societies don't go through menopause.

THAT'S ANOLE

Anoles are arboreal lizards capable of climbing up vertical surfaces, thanks to specialized toe pads. There are more than 375 species in the world, but only one is native to the United States (although a few nonnative species are established in Florida). Green anoles range from East Texas to southern Virginia. They come in avocado shades, from a bright-yellowish tinge to a rich brown. Their ability to change colors is poorly understood, but camouflage doesn't appear to be the primary driver. Instead it may be related to social displays and communications. Males sport a pinkish flap of skin on their throats called a dewlap, and they will perform elaborate push-up displays to impress females and to intimidate rival males. Females lay a single egg every couple of weeks throughout the breeding season, depositing fifteen or so eggs annually.

Green Anole

Introduced to Hawaiʻi in the 1950s, green anoles are also now invasive on the Ogasawara Islands in Japan, where they are responsible for driving multiple native species to extinction.

SPECIAL SOIL

Loess is windswept, silty soil deposited after the last glacial period. These deposits make up unique geologic formations found mainly in China and the American Midwest, although there are other scattered pockets in places like western Mississippi (deposits that originated farther upstream). Structurally, the soil clings to itself when dry, though it will cleave or split at a 90-degree angle. The characteristics of the soil led to the formation of steep sloped loess hills. Vertical bluffs can also be a trait of loess.

The largest deposit of loess stretches along the Missouri River in western Iowa and Missouri and eastern Nebraska and Kansas. Hikes in Loess Bluffs National Wildlife Refuge in Missouri and Loess Hills State Forest in Iowa provide opportunities to witness this exciting dirt firsthand. Restoration work is ongoing to support the prairie hill ridges, which sustain diverse ecosystems and upward of 350 species of plants.

GREEN (JAY) WITH ENVY

South Texas is like no other place in the United States. It represents the far northern reaches for numerous species of plants and wildlife. Ecologically the region is more like Mexico than Amarillo or even San Antonio. Many of the birds and butterflies (see August 28) here seem almost tropical. Green jays play the lead in a vibrant cast of characters. They are common at feeders, but unlike some of the jays farther north, they are never underappreciated. At least not by visitors to the Rio Grande Valley. Great kiskadees play a noble second fiddle in the southern symphony. Nearly as bright, kiskadees have a rich lemon color with a bold call to match, and they squawk conspicuously on open perches. These large flycatchers also visit fruit feeding stations.

Another specialty for the valley is plain chachalaca. Though they could be mistaken for especially scrawny turkeys, they are related to the guans and curassows of Central and South America. Look for them crawling through the trees or scampering along on the ground. You may even hear them raucously calling out "chac a lak, chac a lak, chac a lak."

THE FACE OF A STAR

With twenty-two fleshy tentacles on its face, the star-nosed mole looks like something a child invented, but it is a real species. Fairly widespread, it ranges from eastern Canada south to central Florida. Their face decorations, highly sensitive sensory organs, are covered with twenty-five thousand touch receptors, known as Eimer's organs, which help the animal feel its surroundings. They use their oversized front feet as burrowing shovels and swimming paddles. The species is usually found in moist soils, though some star-nosed moles feed heavily in aquatic environments, where they blow bubbles and then quickly inhale them, allowing the mammals to sniff underwater. This action helps them target mollusks, worms, amphibians, and small

NOVEMBER 10, 1978
Badlands National Park is established in western South Dakota.

Star-nosed Mole

fish. Although other mole species lack the star noses, they also have Eimer's organs; because moles are basically blind, the receptors help them forage for insects, worms, and other invertebrates. In winter, mole tails swell in size and function as fat storage units, important to their survival since these animals don't hibernate.

▶ NOVEMBER 11 ◀
R VERSUS *K*

In ecology, as first proposed by noted researchers Robert MacArthur and E. O. Wilson, there are two different general reproductive strategies recognized among species: the *r*-selected and the *K*-selected species types. The *r*-selected species mature quickly, have short gestations, reproduce quickly, and die relatively young. Examples of *r*-selected species are what people would often call pests or weeds, like rodents, insects, and thistles. These kinds of organisms quickly colonize an area but are susceptible to peaks and crashes. The populations of these *r* species are regulated by maximum reproductive capacity (*r* in population growth equations).

The *K*-selected critters, by contrast, are long-lived and slow to reproduce. They may also provide parental care. Think of things like black bears, American robins, and giant sequoias. These organisms generally maintain fairly stable populations, right around the carrying capacity (*K* in population growth equations) for the environment.

▶ NOVEMBER 12 ◀
MEET THE MURIES

Margaret "Mardy" Murie and her husband, Olaus, are fondly remembered as the protectors of Alaska. The couple met in Fairbanks and married in 1924. They spent their honeymoon trekking five hundred miles by boat and dogsled through the Koyukuk region as part of Olaus's caribou research. Together, they would lead many more expeditions, including a notable journey into the upper Sheenjek Valley, where they gathered

NOVEMBER 12, 1996
Tallgrass Prairie National Preserve is created in Kansas. It is one of the only units of the National Park Service dedicated to this increasingly rare ecosystem.

the wildlife data needed to convince President Dwight Eisenhower to protect the region as the Arctic National Wildlife Range. After her husband's death in 1963, Mardy remained committed to his famed preservation work.

In the late seventies, Mardy testified before Congress on her work as part of a special task force that identified Alaskan land to be considered for federal preservation. In 1980, President Jimmy

Hoodoos

Carter signed the greatest preservation act in American history, the Alaska National Interest Lands Conservation Act. This act protected more than 100 million acres of wilderness, including expanding the size of the Arctic National Wildlife Range to nineteen million acres and changing the name to Arctic National Wildlife Refuge. The Murie Ranch, along the border of Grand Teton National Park near Moose, Wyoming, served as a base camp for conservation for decades. The cabin is listed as a National Historic Landmark and continues to serve as a hub of inspiration.

HOODOOS

Towering totem-like tubes of rock are termed hoodoos, and Bryce Canyon National Park in southern Utah is home to a vast expanse of these geological wonders. Hoodoos form through a three-step process. In the Bryce Canyon region, the process began 50 million years ago with the deposition of rock particles. The layers of rocks included limestones, dolostones, mudstones, siltstone, and sandstones. These layers settled in the floodplain basin of ancient Lake Claron. Geological uplift of the lands occurred as the Farallon and North American tectonic plates collided. Weathering and erosion then sculpted the formations. Ice and rain are the forces most at play in Bryce Canyon, which is at an elevation of over 9,000 feet; since the region experiences a great deal of nightly and seasonal freezing and thawing, it erodes at an accelerated rate. Slightly acidic rainwater helps dissolve the mineral calcium carbonate in the rocks, which also continually sculpts the landscape.

Hoodoos are found on all continents, but the sheer magnitude of these formations in Bryce Canyon is spectacular. Regular snowfall also adds a layer of contrasting white to the rich-red region.

LEGS FOR MILES

It's easy to lump all the leggy invertebrates together, but they are really quite different. Not all bugs are even insects. The "pedes," for example, are more closely related to lobsters, crayfish, and shrimp than they are to insects. Millipedes are decomposers. They don't have a million legs, or even a thousand. They have only two pairs of legs per body segment, and they march along in a straight line. If harassed, they can exude an irritating compound. Millipedes will seek shelter in buildings for the winter season.

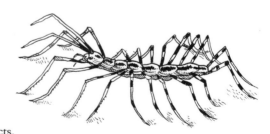

House Centipede

Centipedes, by contrast, are predators, clamping down on insect prey with strong venomous jaws. Each body segment has a single pair of legs, and when scurrying about, centipede bodies undulate from side to side. Most centipedes rarely enter human habitations. House centipedes, however, are a different critter altogether. They are harmless, but their long thin legs can look daunting. They live year-round in homes, where they feed nocturnally on invertebrate prey.

MARSH MATTERS

Although a number of rushes, tules, and reeds serve a similar ecological role, cattails are oftentimes synonymous with freshwater marsh wetlands. Cattails are found throughout the world, especially in the Northern Hemisphere. Both native and nonnative species (and hybrids) are seen in North America. The plants go dormant in winter, but they can still be important habitats because stalks accumulate over the years. Pheasant and deer hunters know that the layered structure of a cattail marsh provides good cover and can create thermal microclimates where wildlife can hunker down. Come summer, everything from bitterns to blackbirds move in. Cattails can be too much of a good thing, though, since under some conditions, they form monoculture stands and choke out species diversity in a wetland.

Many segments of cattails, from roots to stems, are consumed by people. The pollen has been used as a flour substitute and a thickening agent. The fluff has also been used as a stuffing for pillows and mattresses. Highly flammable, the plant's pollen is also a go-to firestarter and is sometimes used in fireworks.

Cattails

FIRST CLASS STAMP

A single postage stamp can mail an envelope across the country, but a different stamp has an even more lasting impact. All hunters of migratory waterfowl, including ducks and geese, must buy the Federal Duck Stamp, officially the Migratory Bird Hunting and Conservation Stamp. Increasingly, nonhunters also purchase the stamp as a way to support conservation. Nearly all of the proceeds from the sales of the stamp directly conserve wildlife habitat, especially within the national wildlife refuge system. To date, more than $1 billion has helped preserve 6 million acres of habitat. As a bonus, the stamp serves as an entry pass for many refuges.

When the Federal Duck Stamp program was initiated in 1934, it featured artwork by J. N. "Ding" Darling (see October 14), a noted wildlife cartoonist, director of the Bureau of Biological Survey, and the man who started the program. Today, a panel of experts chooses the duck stamp designs in one of the most highly prestigious wildlife art contests in the world. A Junior Duck Stamp Conservation program couples an art competition with a conservation-themed message component. The proceeds from this effort help fund environmental educational programming.

▶ NOVEMBER 17 ◀

LIP CURLS

In photos, bull elk and bighorn sheep rams are often shown with outstretched necks and curled upper lips that would put Elvis to shame, a pose related to the breeding season rut. These elk and sheep have a special sensory organ, the vomeronasal, or Jacobson's, organ located above the roof of their mouth at the base of their nasal cavity. They curl their lip to use this organ to collect pheromones and other scents. Female animals also employ this flehmen response, especially after giving birth, and young animals use the behavior as they become aware of their surroundings. The vomeronasal organ is separate from the olfactory system, instead interacting with the hypothalamus region of the brain.

Bighorn Sheep

The flehmen response allows the animal to sense information left behind in the environment. Cats scent-mark territory, and other individuals can use their vomeronasal organ to detect this scent, for example.

▶ NOVEMBER 18 ◀

DELTA DEPOSITS

Water is one of the most powerful forces on the planet. Without the flowing Colorado River, there'd be no Grand Canyon. But where does all of that carved-out soil end up? Rivers carry eroded sediments downstream, and much of this material makes the journey all the way to the coasts. Water flows more slowly as it enters the mouth of a river, so sediment is deposited and forms a delta. The term *delta* refers to the triangle-shaped accumulation, a feature that resembles the Greek letter of the same name. Similar formations, alluvial fans, are the deposits from seasonal waterways and are often located along mountain bases.

Natural erosion accounts for roughly 30 percent of the sediment in waterways of the US, and the rest comes from land use practices. The runoff includes soils, pet waste, road salt, pesticides, and fertilizers—anything really. "We all live downstream," as the saying goes, but equally as important, we all also live upstream. What we do in our backyards affects the oceans, no matter how far inland we live.

NOVEMBER 19

FALLING LEAVES

Deciduous trees prepare for winter by shedding their leaves. Summer leaves, green with chlorophyll, photosynthesize to provide nutrients to the trees. Then in the fall, as daylight shortens and temperatures drop, the leaves lose their chlorophyll, exposing yellows, oranges, and reds. Leaves dislodge from the branch at the abscission zone. These specialized cells help anchor the leaves in place throughout the summer, but in autumn, these cell walls get thinner and thinner. Eventually they completely separate, and the leaves flutter to the ground. Molecules of lignin and suberin build at the junction between twig and leaf petiole, helping protect the tree from pathogens and water loss at these gaps.

> **NOVEMBER 19, 1919**
>
> Zion National Park is established on this date. The core of Zion Canyon, in southwest Utah, was initially recognized as Mukuntuweap National Monument in 1909, but the boundaries were expanded and the designation was changed to Zion National Monument in 1918. And then a year later it became a national park.

Fallen leaves accumulate on the forest floor, providing an insulating layer for plants and invertebrates, and while it's tempting to rake them up in your yard, leaving them can also be beneficial. In the spring, increased daylight and warmer temperatures lead trees to leaf out and the cycle repeats for another year.

NOVEMBER 20

LET'S GET BLITZED

There are many ways to enjoy nature. From passive appreciation to fully quantifiable, all of them are correct. You don't have to be able to identify anything in nature to enjoy it. Just being outside in the fresh air can be relaxing. For some people, nature immersion therapy means hiking the Appalachian Trail; for others, walking the dog around the block is perfect. Some folks find that photographing or tallying what they see is the most satisfying aspect. In the birding world, this quantification can be taken to the extreme; for birders, for instance, a Big Year means trying to see as many species as possible in one year.

Another form of list-making extends observation beyond the birds. A BioBlitz is a coordinated effort to document all of the living things in a particular location. Panels of experts are usually brought in to oversee identification, but volunteers are also essential to the event. National Geographic maintains a partial list of BioBlitzes throughout the country (www.nationalgeographic.org/projects/bioblitz). Many local nature organizations host their own tallies too.

NOVEMBER 21
BIRD TOES

Next time you see a bird, pay particular attention to its feet. The vast majority of birds have four toes on each foot, three facing forward plus one facing backward, known as the hallux. This anisodactyl foot arrangement is common in perching songbirds, including nuthatches, but a few familiar backyard visitors show a different arrangement. To aid in tree climbing, woodpeckers have zygodactyl feet with two toes in the front and two in the back (exceptions include three-toed and black-backed woodpeckers). Members of the cuckoo family, including roadrunners, share this zygodactyl arrangement. Owls and osprey can pivot a toe from the front to the back when grasping prey species.

White Breasted Nuthatch

It's more difficult to observe the feet of waterbirds, but you're probably aware that ducks and geese have webbed feet (see November 4). Pelicans and cormorants have webbing between four toes, giving them totipalmate feet. Grebes, coots, and phalaropes have partially lobed toes. Perhaps the most specialized of toes are pectinate toes. These comblike appendages aid in preening and are found in some of the herons, egrets, bitterns, barn owls, and the nightjars, like nighthawks and whip-poor-wills.

NOVEMBER 22
WATER WONDERS: ICE AND SNOW

Water is a magical substance. When frozen, water expands, but this isn't the only reason why ice floats. Water is most dense at 39.2 degrees Fahrenheit. Above and below this temperature, water is less dense. This density gradient eventually leads to colder (and less dense) surface water that freezes, which is why lakes are covered by sheets of ice instead of freezing from the bottom up.

The properties of water impact snow too. It is more difficult to make snowballs out of drier, fluffier snow. These are not the right conditions for elastic snow either. Individual flakes amass, accumulating as drifts, but the moisture level throughout the drift isn't uniform and is constantly changing. Melting, freezing, and compaction all play a role, and eventually the grain size of the snowpack becomes relatively uniform and stable. When snow has a more uniform grain size, it becomes stronger and can bend around and overhang objects.

▶ NOVEMBER 23 ◀

BEARDED INVASION

Yellowstone and Grand Teton National Parks and surrounding national forests are home to some of the largest concentrations of wildlife in the world. Millions of visitors travel from around the globe to spot the megafauna of the region. But one species is increasing in number and wearing out its welcome: mountain goats, which aren't native to the ecosystem. These shaggy white beasts were relocated to southern Montana in the 1940s and 1950s. By the 1990s the species had established a foothold in the northern reaches of Yellowstone. A similar story is playing out in the Tetons in Wyoming. Mountain goats were released in Idaho in the 1960s and 1970s. These herds have expanded into Grand Teton National Park, and potentially pose a risk to the area's native bighorn sheep. Not only are the two species competing directly for resources, but disease transmission is a concern of wildlife biologists. The fragile alpine environments are also at risk from these bearded invaders.

Washington state is home to the highest numbers of mountain goats in their native range, although within the state, the species has existed on the Olympic Peninsula since they were introduced in the 1920s. The National Park Service is relocating the goats to the North Cascades, where they are native. In the Pacific Northwest, November is breeding season for the goats, but hikers tend to spot them more often in summer.

▶ NOVEMBER 24 ◀

FLY AGARIC

A classic red mushroom with flakes of white, *Amanita muscaria* is considered poisonous and can cause death. Some say that the hallucinogenic properties of this species are perhaps the origins of the story of the flying reindeer that haul Santa around the globe in December. Even if you don't plan on taking a ride with Mr. Claus, the amanita mushroom is worth appreciating in the wild. It is sometimes called the fly agaric, a name that harkens back to at least the 1200s, when the species was added to milk and used as an insecticide to repel flies. The species first appears aboveground looking like a little white egg. The white spots on more mature specimens are remnant speckles of the universal veil that encases the entire mushroom during its early stages.

Initially native to the Northern Hemisphere, amanita is a cosmopolitan species that has been introduced worldwide, usually in forest plantation situations. Fruiting bodies of the fungus can appear in summer or fall throughout much of North America. It tends to grow later in the season in the Pacific Northwest.

▶ NOVEMBER 25 ◀

GOBBLE DAY

The origins of the traditional Thanksgiving turkey are a bit vague. At the first Thanksgiving in 1621, it seems that ducks and geese were the main fowl for the feast, although turkey might have been served. Venison was central to the meal, as were fish and shellfish. Cranberries might have been eaten, but not in sauce form and certainly not out of a can. Squash, perhaps even pumpkin, was likely on the table—sadly, though, not as pie. Turkey as the main dish didn't take off until the mid-1800s. The vast majority of holiday turkey eaten today is farm-raised, although hunters still bag gobblers from coast to coast.

Alaska is the only state without wild turkeys, but the species wasn't always so abundant. Habitat loss and market hunting led to declines and extirpations of the species in the 1800s. Fewer than 250,000 turkeys remained by the 1930s. Much habitat work and the relocation of birds reversed the decline. Now over 6 million wild turkeys roam free.

▶ NOVEMBER 26 ◀

EIGHT-FOOTED SWIMMERS

Octopodidae, the name for the octopus family, literally means "eight-footed." Though their eight tentacles help define what it means to be an octopus, there is much more to these fascinating swimmers. For starters, each of those eight legs has a separate brain, in addition to the creature's central brain. An octopus has three hearts: two that supply blood to the gills and a third that circulates blood throughout the rest of the body.

They are notorious for escaping aquariums in captivity since they can essentially squeeze through whatever their beak fits through. Regarded as highly intelligent, they've been recorded opening screw-capped jars. Octopuses have more neurons in their tentacles than in their central brains.

The giant Pacific octopus is the largest species. Most weigh about one hundred pounds and measure sixteen feet long or more. The largest recorded specimen was twice as long and weighed a whopping six hundred pounds. Normally a reddish hue, giant Pacific octopus skin can change colors thanks to specialized chromatophore pigment cells. Camouflage is this cephalopod's first line of defense. Releasing clouds of black ink and a quick exit is the last resort for an octopus in danger.

NOVEMBER 27

KEYSTONE SPECIES

Like the keystone supporting an arch, keystone species are any critters that have an exaggerated impact on their ecosystem. If they disappeared, the entire landscape would be impacted. The concept was first applied to sea stars for their ability to alter rocky intertidal zones. Sea otters are another iconic example for the way they impact kelp beds and change the ecosystem. Beavers are also classic keystone species. By damming up flowing water, they create ponds of slower-moving water, fundamentally altering the habitat. In many streams, these deeper pools provide nursery habitat for fish, amphibians, and invertebrates. And the impact of beavers on the land goes beyond chewing down some trees; their actions shift the canopy structure and can reset succession.

Similarly, prairie dogs impact grasslands, creating an entire ecosystem of prairie plants and animals. Numerous creatures, from black widow spiders to burrowing owls, utilize prairie dog burrows. Vegetative changes occur on the prairie dog towns as well, and while biomass decreases, many herbivores selectively graze on these pockets of habitat. A true keystone species, prairie dogs are also prey for numerous grassland predators, including one of the most endangered mammals in North America, the black-footed ferret (see May 17).

NOVEMBER 28

FAR FROM HOME

There is a familiarity in nature that many people find comforting. With a little research about or experience with an area, you can have a fairly good idea of what birds you should expect in a habitat during any season. But every year unexpected species pop up in unlikely places. Some are blown off track by strong winds. Others seem to migrate the exact opposite direction of their normal pattern. And a few individuals materialize seemingly out of nowhere.

Birds that show up far from their expected ranges are called vagrants. Most of these individuals represent evolutionary dead ends. They aren't likely to reproduce or even survive in these far-flung locales. That doesn't mean people don't get excited about them, though. When species from other continents end up in North America, birders realize it may be the only chance they'll have at seeing that bird. And vagrancy works both directions. The twitchers (extreme birdwatchers of Britain) also

Sea Otter

lose their collective minds when common American birds end up on the other side of the pond. But vagrancy doesn't have to be that extreme. It can also apply to a bird that is a few states over from where you'd normally expect it to be.

THE ROOT BEER TREE

Sassafras

Sassafras is a deciduous tree noted for its aromatic leaves and bark. Originally the bark and roots were used to flavor root beers, although these ingredients have largely been replaced with artificial flavorings thanks to restrictions imposed by the US Food and Drug Administration. The leaves of the tree are ground into filé, a thickening agent preferred for gumbo. Botanically, sassafras is a hardy tree (though it's shrubbier in the North), found from Texas and Oklahoma to southern Ontario. In spring, the plant develops showy yellow-greenish flowers. Summertime brings out the noticeably lobed leaves. Sassafras leaves have been described as mittens, with both left- and right-handed leaves occurring. Three-lobed and no-lobe leaves are also found. The leaves turn a brilliant orange, yellow, or scarlet in fall. The fruits of the sassafras ripen in autumn as well and are an important food for black bear, deer, raccoons, fox, and many bird species, including wild turkey. These foragers help disperse the seeds and assist sassafras as an early successional species.

CATERPILLAR METEOROLOGISTS

How harsh is winter going to be this year? There are plenty of legends on the best ways to predict the weather. Spoiler alert: None of them have much merit. People have long turned to insects to help forecast the cold season. Seeing ants marching in straight lines is a bad weather omen, right? Or if you see a lot of crickets earlier in the fall, winter is close behind—same if spiders are spinning especially large webs in fall. But perhaps the most popular technique is to ponder whether or not to bust out the parka at the sight of a black-and-orange woolly bear or woolly worm caterpillar in the fall.

These fuzzy larvae of the Isabella tiger moth are seeking shelter, no matter the forecast, as they overwinter in caterpillar form. In some regions, more orange is said to be indicative of milder temps, but the opposite coloration is thought to apply elsewhere. While the folklore is fun, don't bank on a caterpillar replacing your local meteorologist.

BEACH BANANAS

Found on both coasts, harbor seals are one of the most common marine mammals in the United States, Canada, and across the Northern Hemisphere. They are seen along the Pacific and northern Atlantic year-round. In winter, some individuals move farther south, making early December prime time for seeing seals in the Mid-Atlantic. Classified as true seals (see September 6), harbor seals lack external ear flaps. With just their heads sticking out of the water, they can look like balloons with eyes, and when they rest on land, they lie in a banana position, keeping their heads and rear flippers up off the ground. Their short forelimb flippers help them undulate on land. Harbor seals have sharp teeth for grasping and tearing fish and flat back teeth for crushing shells and crustaceans.

SEDIMENTARY ROCKS

Of the three main categories of rocks, sedimentary rocks are the most abundant, accounting for 75 percent of the rocks on Earth (see August 11 for igneous rocks and June 8 for metamorphic rocks). Sedimentary rocks form from the erosion and then deposition of particles over time. These sediments accumulate and are compacted before being lithified, or cemented, to form rock. Over time these rocks continue to erode into natural sculptures. Many familiar rocks are sedimentary, including limestone, sandstone, clay, shale, and chalk. They can be found in many locations, but southern Utah has some of the finest sedimentary landscapes. The red rocky landscape from Arches to Zion is a sandstone showcase.

DECEMBER 2, 1980

The Alaska National Interest Lands Conservation Act is signed into law by President Jimmy Carter on this date. It greatly expanded the area of lands administered by the National Park Service in the Last Frontier State.

BONE IN

Many male primates—but not humans, mind you—have bones in their penises called baculum. So do various members of the Carnivora order, from raccoons to walruses. Some rodents, shrews, and bats do too. Evolutionarily, the baculum, or os penis, has arisen in at least nine different families. It has also disappeared from ten different lineages of mammals. This wide-ranging trait is thought to serve a couple of different functions. For some species, the purpose seems to be to extend the duration of copulation. For others, especially the bats, in addition to support, the baculum may protect the urethra from compression.

Baculum aren't the only curiosity of the nether regions of nature. Opossums have bifurcated (forked) penises, which helps explain the false notion that they breed via the nostrils of females. The split tip aligns with the lateral vaginal canals of the species. And while most birds breed with a cloacal kiss, ducks have penises, some of which are absurdly complicated in shape. Think corkscrews.

THE SILKY-FLYCATCHER

Resembling a thin, jet-black or slate-gray cardinal, the phainopepla is in a genus all its own. The southwestern species is related to the silky-flycatchers of Central and South America. Both sexes flash white wing feathers and sport ruby-red eyes, and the males are darker in color than the females. Nesting for phainopepla is curious. The birds are territorial during a late-winter to early-spring breeding season in the desert lowlands. But later in the season, they are loosely colonial for an additional breeding season in cooler oak and sycamore canyons. Phainopeplas overwinter in desert lowlands or push south into Mexico.

Throughout the year, mistletoe berries are an important component of the phainopepla diet. Individual birds can consume upward of a thousand berries each day. The gizzards of the species are specialized to aid in berry digestion. True to their silky-flycatcher family, they also eat aerial insects. A capable mimic, the species has been heard imitating a dozen other birds.

Phainopepla

DECEMBER 5

AMERICAN WOODSMAN

As an eighteen-year-old Frenchman, John James Audubon moved to his family's property at Mill Grove, Pennsylvania, in 1803. This is where he was the first to band a bird in North America (see October 17). After a stint at Mill Grove, Audubon, along with his neighobr and future wife Lucy, relocated to western Kentucky to operate a dry-goods store. The store hit hard times, and Audubon was briefly jailed for bankruptcy in 1819. Upon his release from jail, while Lucy earned wages as a tutor, John James toured the continent collecting specimens and depicting birds and their habitats. Like Alexander Wilson before him, Audubon's goal was not only to document all of the North American bird species but also to pay special attention to the habitats they were found in. Wilson's *American Ornithology* was published in America, while Audubon found early success with *Birds of America* in Europe, where he was known as the American Woodsman.

John James Audubon, along with his son, also published a book on mammals, *Viviparous Quadrupeds of North America* but his name will always be linked with birds. Conservation leader George Bird Grinnell was tutored by none other than Lucy Audubon, so when he became one of the early founders of an organization dedicated to birds and habitat, he chose Audubon, a fitting name.

DECEMBER 6

MIGHTY RIVERS START SMALL

Even the mightiest of rivers starts out as a small trickle of water somewhere far from its mouth. In the case of the Mississippi, its headwaters are found, at least symbolically, at Lake Itasca in Itasca State Park in northern Minnesota. A natural spring keeps the water mostly free of ice. As winter sets in, bald eagles congregate along the upper stretches of the Mississippi River.

A historic marker from the 1930s indicates the total mileage for the Mississippi is 2,552 miles. Channelization has shortened it by a couple of hundred miles, but people still attempt the now 2,300-plus-mile journey down the river each year. From Lake Itasca to sea level, it drops 1,475 feet in elevation. Collecting water from hundreds of tributaries, the Mississippi River watershed drains 41 percent of the United States.

RIVER OF GRASS

A different kind of river flows through Florida, one described by journalist and conservationist Marjory Stoneman Douglas in her 1947 book, *The Everglades: River of Grass*. The book was published the same year Everglades National Park was established, on December 7. The glades are open swaths of sawgrass, and historically a wide sheetflow of water moved slowly and freely from Lake Okeechobee to the Florida Bay. This wide, uninterrupted swath of water is no longer. The hydrology of the area has been greatly impacted by levees and dikes and is now heavily managed. Restoration efforts are ongoing to return the flows to the Everglades.

The national park is far more than sawgrass, though. The subtropical wilderness contains nine distinct habitat types. While the Shark Valley observation tower will put you in the heart of the "river of grass," one of the best ways to explore the park is by canoe along the ninety-nine-mile Wilderness Waterway route.

EVERGLADES PYTHONS

Everglades National Park and adjacent Big Cypress National Preserve are home to delicate ecosystems. A 1992 natural disaster led to an ecological battle that tipped the balance: When Hurricane Andrew hit South Florida, some of the destruction was immediately apparent. Other consequences are playing out over a longer time frame. For example, the destruction of a snake-breeding facility led to a giant predator being released into the habitats of the region. Burmese pythons, native to Southeast Asia, can grow to be longer than seventeen feet from tongue to tail, although most are more like eight to ten feet long. They basically reach full size by their fourth year.

In the years following Hurricane Andrew, local populations of deer, raccoon, opossum, fox, rabbit, and wading birds have all been decimated in areas where the pythons dominate. Trained professionals have removed thousands of pythons throughout the region, but the population of snakes continues to expand, and the impacts to the ecosystem continue to be felt.

Burmese python

THE BEARDED ONE

Musk ox are named for a strong odor that the males release during the breeding rut. The Inuit name, *umingmak*, translates to "bearded one," and the genus, *Ovibos*, means "sheep ox" in Latin. Found at the northern extremes of the planet, musk ox are thought to have moved into North America from Siberia between 200,000 and 90,000 years ago.

When food is sparse in winter, herds of musk ox sniff out dried grasses, sedges, willows, and lichen and use their hooves and heads to dig through the snow. Musk ox are a herding species, which they use to their advantage when it comes to avoiding predators. A group of musk ox will circle up, forming a ring of horns. Both males and females sport hefty horns.

Dense fur is another characteristic of musk ox, with both long guard hairs and thicker inner wool called qiviut. This natural fiber can be spun into expensive yarn, noted for its soft qualities.

GEMINIDS

The Geminid meteor shower is one of the most reliable of the year. It peaks in early to mid-December and consistently produces good numbers of meteors for several nights before and after. The burning debris responsible for the streaks of light comes from 3200 Phaethon, an asteroid. Watch for the meteors in the southeastern sky, near the Gemini constellation, especially radiating from the direction of the star Castor. Viewing improves as the constellation rises higher in the sky. The meteors reach their pinnacle by two o'clock in the morning. Don't fret if you can't rally that late—you have a good chance of seeing them earlier too. Rates can reach 160 meteors per hour, although not all of these will be visible.

The most important factor in viewing any meteor shower is a very dark sky. (To learn about efforts to lessen light pollution, see April 1.) Try to get far away from the city lights to start looking up.

THE SAME CRITTERS BUT DIFFERENT

An ecological principle described by German biologist Carl Bergmann in 1847 holds that widespread species are larger in colder northern regions and smaller in warmer areas. There are some exceptions to the so-called Bergmann's rule, but the sliding scale of sizes applies to numerous critters across taxa. Some thirty years later, Joel Asaph Allen developed a similar rule correlating surface-area-to-volume ratios (SAV) of endotherms (warm-blooded animals) with climates. Heavier bodied coyotes of the North have relatively shorter legs and smaller ears, while southern canids tend to have longer, lankier limbs and larger ears, for example. (It's easy to see why visitors from the South, for example, mistake coyotes in Yellowstone for wolves.)

DECEMBER 11

International Mountain Day is a United Nations celebration highlighting the vital roles mountains play in the lives of many.

These varying morphologies are explained by thermoregulation needs. Lower SAV animals retain body heat more efficiently. In hot climates, high SAV ratios make it easier to cool off. Antelope jackrabbit ears act like air-conditioning in the South, while related snowshoe hares have stubbier ear stalks to combat frostbite.

> ## DECEMBER 12 ◀

SMOOTH (SUMAC) OPERATOR

Smooth sumac is the only species of tree or shrub native to all forty-eight contiguous states. The ten-to-twenty-foot-tall bush grows in dense colonies. Common along the edge of fields, it's also roadside scenery along many routes. This time of year the foliage is especially showy with bright red, yellow, and orange. These compound leaves look like multiple leaves sprouting from a single stem. Smooth sumac and related species, including shining and staghorn, have seed-bearing drupes, a kind of fruit, that are fuzzy, red, and grow in a cone-like shape, making them look like candelabras.

A related species of sumac is especially popular in Middle Eastern cooking, where it's ground and sprinkled on pitas or hummus, and its tart flavor is the backbone to za'atar spice blends. The drupes of the North American plant are said to have a slightly citrusy flavor and can be used to make a tea or lemonade alternative. The stems of sumac are easily hollowed out and have been fashioned into pipe stems and spiles for channeling maple sap from tapped trees.

Smooth Sumac

WINTER TROGLODYTE

As you may expect from a bird family named Troglodytidae, wrens explore the nooks and crannies of the world. They tend to be boisterous callers, so their presence is rarely a secret. There are more than eighty species found throughout the Americas, but just one species, the European wren, is found in the rest of the world. In the colder months, winter wrens are a widespread but hard to find species in much of the eastern United States. Pacific wrens (formerly classified with winter wren) occupy the West. The Carolina wren, historically a southeastern bird, has been expanding its range northward for decades, and cold weather occasionally wipes out the birds in the upper reaches of this range expansion. A number of species, including canyon, rock, Bewick's, and cactus wren, inhabit western states. Marsh and sedge wrens can be found in specific habitats. Stretching from southern South America to Canada, the house wren has one of the broadest breeding ranges of any songbird species, but in winter its distribution is limited to the southern tier of states in the US.

EAB

Few people embrace bitter cold, but there may be an upside to the low temperatures in the wild as extended cold periods can impact overwintering insects. Since its discovery in the United States in 2002, the emerald ash borer (EAB) has exploded throughout the East and Midwest. This nonnative invasive invertebrate is wreaking havoc on forests, one tree at a time. As opposed to the ash trees of Asia, where the emerald ash borer is native, the species of ash trees of North America did not evolve with this insect, so larvae kills the trees as it girdles them. The iridescent metallic-green beetle is about half an inch long as an adult. In summer, adults lay eggs on the bark of ash trees, then the hatched larvae tunnel under the bark, creating S-shaped tunnels as they feed, and they overwinter in larval form in the inner tree. The insect tolerates cold to an extent, but frigid temperatures can kill up to 80 percent of populations.

Unfortunately, winter die-offs aren't going to eliminate the threat of the EAB. The surviving larvae emerge as adults in the spring and begin to repopulate the area. While native species can also feel the adverse effects of a cold snap, in most instances, they are more adapted to tolerate winter conditions.

TEA WATER

Turbidity is the measure of how clear water is. To measure turbidity, you can lower a black-and-white Secchi disk into water until you can no longer see it. The water's transparency, or lack thereof, can be an indicator of water quality. Higher levels of turbidity, meaning water that is less clear, can cause higher temperatures, lower oxygen, and decreased photosynthesis. Water can be murkier in certain seasons or due to suspended sediment or algal blooms. And though murky water can be brown, there is a difference between murky water and tannic (clear brown) water.

Tannic water can be found in a variety of habitats, although it's often associated with southern swamps and northern bogs. This tea-stained effect is caused by tannin chemicals commonly found in plants, especially conifer trees, deciduous leaves, tree bark, and a variety of flowering plants. The word itself comes from *tanna*, the Old German word for oak. The astringent flavor of your favorite red wine may come from the tannin in the oak barrels the beverage was aged in.

TANNING WITH TANNINS

Humans have used tannins, the compounds that stain water brown, for years. The astringent qualities found in tannins are what made witch hazel and cranberries go-to cures historically. As far back as 5000 BC, the Egyptians, Greeks, Romans, and Chinese were all tanning animal hides using concoctions of liqueurs derived from animal brains and plant tannins. Oak, willow, sumac, maple, eucalyptus, and red mangrove have all been used this way. For the plants, the tannins serve multiple functions, including protection from bacteria, fungal infection, and insect pests. Many young plants show elevated levels of tannins, which is thought to help give them a chance to grow before being consumed. Beavers selectively harvest cottonwoods with lower levels of tannins. Underripe fruits are also high in tannins, which keeps animals from eating the fruits before the seeds are mature and prepared for dispersal.

CARTWHEELS OF LOVE

Bald eagles form pair bonds that last from year to year. Courtship and pair bond renewal, including perching together, mutual preening and bill touching, and collaborating on nest construction, occur in winter as the eagles prepare for nesting season. Their courtship culminates in elaborate sky dancing,

Eagle Courtship

which entails high-speed chases, riding thermals, and soaring together. The most dramatic move is when the pair clasps talons and cartwheels downward, releasing as they approach the ground. The cloacal kiss of copulation follows, with egg-laying five to ten days later.

Eagle nests can be some of the largest structures in the avian world. Each year, the pair will rebuild if need be. Alternatively, they can add to an existing nest year after year. They use sticks to strengthen the platform and grassy materials to make a soft cup for the eggs and nestlings. The largest documented bald eagle nest was built near Saint Petersburg, Florida, in the 1960s. It was nearly ten feet wide, twenty feet deep, and weighed an estimated 4,400 pounds.

> DECEMBER 18 <

RIBBONS OF INDIGO

Indigos are nearly black snakes, glossy and sleek in appearance. They reach lengths of seven feet, making them among the largest of the North American snakes. Though they are nonvenomous and nonconstricting, indigo snakes are still impressive diurnal hunters. They rely on speed to take down a variety of prey species with a strong bite. The species is notorious for taking out copperheads, and they can tolerate rattlesnake venom. They will eat anything that becomes readily available—rodents, rabbits, birds, frogs, turtles. Breeding occurs in winter to early spring, with the nearly two-foot-long hatchlings emerging from eggs in spring or early summer.

All indigo snake species are threatened across their ranges, both in South Texas and in the Southeast. After having been extirpated from Alabama, they are now being reintroduced in Conecuh National Forest, in an effort to restore this top predator to the longleaf pine ecosystem.

> DECEMBER 19 <

ALOHA BIRDS

As winter sets in across much of the mainland US, the islands of Hawai'i beckon. The birdlife of this isolated paradise is rife with endemic species (see October 8), species that are found nowhere else. The state bird is the endangered nēnē, or Hawaiian goose. In the same genus as Canada geese, these birds have evolved longer legs, smaller bodies, and less webbing in their feet.

DECEMBER 19, 2014
Tule Springs Fossil Beds National Monument, a significant archaeological site containing fossils of mammoth, lion, and camel, is established in Nevada.

They are capable of flight, but nēnē spend considerable time walking on the terrain. Captive breeding programs have helped populations of the goose rebound.

A number of other birds add to the state's diversity, including the honeycreepers, although many of them have gone extinct or are endangered. The brilliantly bright 'i'iwi, or scarlet honeycreeper, has a curved orange bill for sipping nectar from the local lobelia flowers. And the yellow-green-hued 'amakihi, another honeycreeper, feeds on insects, fruits, and nectar with a piercing bill. The pueo, or Hawaiian owl, is a doppelgänger of the short-eared owl. An ivory bird with contrasting black eyes and bill is the white tern, or manu-o-Kū. Also called the fairy tern, this urban dweller has been declared the official bird of Honolulu. Unfortunately, the birds of Hawai'i face pressures of nonnative disease, habitat loss, and introduced predators, such as the mongoose.

▶ DECEMBER 20 ◀
DECK THE HALLS

Decorating with boughs of holly goes back centuries, and the practice has long been incorporated into various religious traditions. Roman, Celt, and Norse pagans all celebrated the shortest day of the year by hanging a number of evergreen branches, including holly, ivy, bay, fir, rosemary, laurel, boxwood, and mistletoe. For Romans celebrating Saturnalia (a festival to honor the god Saturn), holly represented Saturn, the god of agriculture. In Celtic mythology, the Holly King and the Oak King battled at midsummer and midwinter, the victor bringing in the annual change of seasons. And druids believed the spiky leaves of holly would help ward off evil spirits. For Christians, the holly branches represent a crown of thorns, with the red berries portraying droplets of blood.

The vibrant-red poisonous bulbs of holly aren't berries but instead are a kind of fruit called drupes, akin to olives, cherries, and peaches. Holly trees are either male or female, but only females have the drupes. Botanically, there are hundreds of holly species, with fifteen types identified in North America. The most common, American holly historically was widely harvested for holiday decoration. Delaware led the way in commercial harvest, and the state declared American holly its official state tree in 1939. These days, artificial decorations have almost completely replaced fresh boughs.

American Holly

KEY TO IMPORTANT SPECIES, IDEAS, PLACES, AND MORE

ALGAE

algae (April 21, July 4, August 8, December 23)
dinoflagellates (July 4)
phytoplankton (January 16)
seaweed (July 9)
watermelon snow (April 21)
zooplankton (January 16)

AMPHIBIANS

American bullfrog (May 8)
eastern hellbender (April 9)
eastern narrow-mouthed toad (February 21,
    March 14)
salamanders (April 9, May 8)
spring peeper (February 6)
western narrow-mouthed toad (February 21)
wood frogs (January 6, February 6)

AQUATIC

aquifer (September 19)
bayou (February 7)
delta (November 18)
Ogallala Aquifer (September 19)
oxbow (May 27)
saltwater lake (August 8)
tides (September 27)

ASTRONOMY

Alnilam (January 2)
Alnitak (January 2)
Andromeda Galaxy (April 2)
Aurora Borealis (November 3)
Autumn Equinox (September 21)
Betelgeuse (January 2)
Big Dipper (June 1)
comets (March 2)
dwarf planet (October 3)
Dubhe (June 1)
Earth's axis (June 21, September 21)
eclipse (May 4, September 1)
Gemini (February 3)

Geminid meteor shower (December 10)
Gregorian calendar (February 29)
Hale-Bopp Comet (March 2)
Halley's Comet (March 2)
Kuiper Belt (March 2, October 3)
Leap Day, Leap Year (February 29)
light pollution (April 1)
Little Dipper (June 1)
Merak (June 1)
meteors (August 12)
Milky Way (April 2)
Mintaka (January 2)
moon (May 4)
North Star (February 3, March 21, June 1)
Northern Hemisphere (March 21, June 21,
    September 21)
northern lights (November 3)
Oort cloud (March 2)
Orion (January 2)
Orion Nebula (January 2)
Perseid meteor shower (August 12)
photoperiod (March 20)
Pluto (October 3)
Polaris (June 1)
Rigel (January 2)
Southern Hemisphere (June 21)
Spring Equinox (March 21)
Summer Solstice (June 21)
Swift-Tuttle Comet (August 12)
Tropic of Cancer (June 21)
Winter Solstice (December 21)

ADAPTATIONS AND BEHAVIORS

allelopathy signals (September 18)
altricial young (May 24)
antlers (December 24, December 30)
baculum (December 3)
bioluminescence (May 29, July 4, October 31)
breeding (January 5, 19, 24, 29; February 18; March
    11, 13, 17, 30; April 3, 16; May 3, 17, 23; June 22;
    July 1, 6; September 10; October 4, 12, 19)
buzz pollinators (June 29)

camouflage (July 28, August 7, September 10, October 9, November 26)

commensalism (April 18)

communal cavity roosting (February 4)

communication among trees (September 18)

competition (April 18)

convergent evolution (August 30)

coprophagy (January 30)

courtship (September 5, October 9, December 17)

crop milk (May 30)

crowns of trees (January 23)

dawn chorus, songbirds (May 19)

dendrochronology (January 18)

digitigrade locomotion (May 12)

echolocation (June 24)

Eimer's organs (November 10)

endemic species (October 8, 18; December 19)

environmental mimicry (September 10)

estivation (January 14, August 31)

extinction (November 5)

extra-pair copulation, birds (June 25)

eye location, related to species (May 31)

feces, eating (January 30)

feeding cessation (February 12)

feet types, birds (November 21)

food caching (January 4, 11; March 14; October 4, 5)

foraging (February 12)

freezing and thawing (January 6, February 6)

gait (January 20)

glycolipids (January 6)

granaries, woodpeckers (October 4)

hibernation (March 7, August 31)

horns (December 30)

hybridization (May 20; June 26, July 6, 10; August 2)

irruptive migration or irruption (January 12, 19)

larder hoarding (October 5)

mandibles (January 19, September 15)

mast crops (September 30)

mating (January 14, September 14, October 12)

menopause (November 6)

metamorphosis (June 11, October 2)

migration (January 12, 15; March 11, 12, 13, 16, 20; April 28; May 3; August 2, 25; September 11, 22; October 1, 11, 16, 19)

molting, fur or feathers (January 22, 29; April 16)

mutualism (April 18)

myrmecochory (April 10, 12)

nest boxes, birds (February 4, May 23)

nest parasites (June 16)

nesting (January 28; December 17, 26)

nictitating membrane (June 3)

overharvesting (October 30)

ovoviviparous (August 7, 27)

owl pellets (May 30)

parasitism (April 18)

parenting (March 17; May 24; June 16, 27; October 9)

patagium (March 1, June 24)

paternity, birds (June 25)

phenology (January 1, June 30)

plantigrade locomotion (May 12)

poaching (October 10)

pollination (January 25; March 4, 22, 31; April 10, 17, 18, 19; May 13, 20; June 24, 29; August 16, 28)

precocial young (May 24)

predation (April 18)

predatory species, eyes (May 31)

predator-prey cycles (January 27, September 30)

prey species, eyes (May 31)

questing, ticks (February 28)

radula (January 14)

regional dialects, songbirds (May 26)

regurgitation (May 30)

reproductive strategies (November 11)

rut (October 12, November 17)

scatter hoarding (October 5)

sesamoid bones (December 24)

shell middens (October 6)

skin shedding or molting (January 8, February 28, August 1)

snowplowing, bison (February 26)

straddle (January 20)

stride (January 20)

taste receptors on feet (June 14)

torpor (March 7, August 31)

tracks (January 20)

tree rings (January 18)

unguligrade locomotion (May 12)

vomeronasal organ (November 17)
white-nose syndrome (June 24)
winnowing (June 22)
yarding, deer (February 9)
zygodactyl feet (August 17)

BIRDS

acorn woodpecker (October 4)
alcids (July 13)
American avocet (August 8)
American dipper (August 13)
American goldfinch (January 29, September 4)
American oystercatcher (August 2)
American white pelican (January 24)
American wigeon (February 7)
American woodcock (February 18)
anhinga (August 15)
Archaeopteryx (August 29)
Arctic tern (October 19)
Atlantic puffin (July 13)
auklets (July 13)
bald eagle (December 6, 17)
Baltimore oriole (April 23, July 10)
blackpoll warbler (October 1)
bar-tailed godwit (August 26)
barred owls (January 5)
black-bellied whistling duck (February 7)
black-capped chickadee (October 2)
black-headed grosbeak (May 16)
black-necked stilt (August 8)
black oystercatcher (August 2)
black skimmer (September 15)
blue grosbeak (May 16)
blue-winged teal (February 7)
boreal owl (July 25)
brown-headed cowbird (June 6, 16)
brown pelican (January 24)
Bullock's oriole (April 23, July 10)
buntings (May 19)
burrowing owl (September 3)
butcher-birds (March 14)
California gull (July 24)
Carolina parakeet (November 5)
Carolina wren (December 13)
Caspian tern (July 1)
Cassia crossbill (January 19)

cinnamon teal (August 8)
clapper rail (August 20)
Clark's grebe (April 16)
colonial waterbirds (April 7)
cormorant (August 15)
dabblers, ducks (November 4)
dark-eyed junco (December 22)
divers, ducks (November 4)
dodo (November 5)
downy woodpecker (October 2)
ducks (November 4)
dusky grouse (April 3, October 16)
eared grebe (August 8)
eastern phoebe (October 16)
eastern screech-owl (January 5)
eastern towhee (July 10)
eggs (June 16)
egrets (April 7)
European starling (March 6)
evening grosbeak (May 16)
feathers, shape and type (July 28)
Federal Duck Stamp program (October 14, November 16)
finches (May 19)
flamingos (June 19)
flyways (February 24)
gadwall (February 7)
great auk (November 5)
great kiskadee (November 9)
great horned owl (January 1, 5)
greater roadrunner (August 17)
green jay (November 9)
green-winged teal (February 7)
grouse (March 30, April 3)
grosbeaks (May 16)
guillemots (July 13)
gulls (April 7, July 6)
heath hen (November 5)
herons (April 7)
Hispaniolan crossbill (January 19)
honeycreepers (December 19)
hooded merganser (June 16)
house finch eye disease (February 13)
hummingbirds (August 30)
ibis (April 7)
'i'iwi, or scarlet honeycreeper (December 19)

woodpeckers (June 3)
wrens (December 13)
yellow-rumped warbler (January 9)

BOTANY

agave (January 25)
ash (December 14)
aspens (September 29)
bald cypress (November 2)
bayberry (January 7)
beargrass (July 30)
berries (December 31)
berry bushes (July 23)
black mangrove (February 16)
bladderwort (May 1)
bloodroot (April 12)
bluebonnets (April 17)
bosque or riparian corridor (January 13)
California palm (January 3)
Cambium, trees (January 18)
carnivorous plants (May 1)
Catawba rhododendron (June 17)
cattails (November 15)
cherrybark oak (October 22)
chicory (July 19)
coast or Pacific redwood (January 14, 23)
Colorado columbine (May 20)
conifers (February 9)
cottonwood (January 13)
dandelion (January 26)
Douglas-fir (June 13)
dwarf mistletoe (December 26)
Eastern pasqueflower, prairie crocus (March 29)
eastern witch hazel (February 8)
elk thistle (September 4)
Equisetum (October 23)
evening primrose (April 19)
fiddleheads, ferns (March 27)
fireweed (August 19)
galls (October 2)
ghost orchid (September 12)
ginseng (October 10)
goldenrod (October 2)
great laurel (June 17)
heart rot (February 10)
heartwood (January 18)

hedge balls or hedge apples (October 15)
hemlock (January 14)
honey mesquite (May 15)
holly (December 20)
horsetail (October 23)
house wren (December 13)
jack pine (June 6)
Joshua tree (February 27)
krummholz trees (July 8)
laurel oak (October 22)
loblolly pine (October 22)
lupine (April 17)
magnolia (March 4)
maple tree (January 1)
mesquite (May le15)
mistletoe (December 4, 26)
monocots (January 3)
nandina or heavenly bamboo (December 31)
nocturnal plants (April 19)
nut trees (September 30)
Ohio buckeye (October 26)
Osage orange (October 15)
organ pipe cactus (April 19, May 13)
overcup oak (October 22)
Ozark witch hazel (February 8)
Pacific rhododendron (June 17)
palo verde (April 29)
Pando grove (September 29)
pawpaws (September 17)
persimmon tree (October 22)
phloem (January 18)
pioneer species (July 31)
pitcher plants (May 1)
Pitcher's thistle (September 4)
pneumatophores (February 16)
poison ivy (July 15)
ponderosa pine (June 4)
red mangrove (February 16)
red oaks (October 5)
rhododendron (June 17)
sabal or cabbage palm (January 3)
saguaro cactus (January 25)
sapwood (January 18)
sassafras (November 29)
sawgrass (December 7)
shooting stars wildflower (June 29)

Sitka spruce (January 14)
skunk cabbage (March 22)
smooth sumac (December 12)
southern magnolia (March 4)
Spanish moss (March 15)
subalpine zone (July 8)
sugar maple (March 5)
sundews (May 1)
super-bloom (February 22)
sweet gum (October 22)
tannins (December 16)
thistles (September 4)
tulips (March 12)
Vaccinium bushes (July 23)
velvet mesquite (May 15)
Venus flytrap (May 1)
wax myrtle (January 7)
white oak (October 5)
wild rice (September 25)
witch hazel (February 8)
xylem, trees (March 5)
Yoshino cherry tree (April 6)

CITIZEN SCIENCE OPPORTUNITIES

Appalachian Trail Seasons program (September 20)
Backyard Worlds: Planet 9 (July 3)
BioBlitz (November 20)
bird banding (October 17)
Budburst (March 28)
Certified Wildlife Habitat program (September 8)
Christmas Bird Count (December 25)
Citizen Science Day (April 13)
eBird (October 28)
FrogWatch USA (May 7)
Great Backyard Bird Count (February 15)
Green Wave (January 25)
iNaturalist (September 20, October 28)
iSeahorse program (October 9)
Journey North (March 12)
Mayfly Watch program (January 25)
Mountain Watch program (September 20)
National Moth Week (July 21)
naturalist (May 28)
Nectar Connections (January 25)
Nature's Notebook (January 25)
Pest Patrol (January 25)

Project FeederWatch (February 15)
Project NestWatch (May 23)
Signs of the Seasons (January 25)
water-quality monitoring (June 18)

FISH

alligator gar (October 27)
American eel (May 18)
American shad (May 18)
anadromous fish (May 18)
blueback herring (May 18)
bowfins (March 17)
catadromous fish (May 18)
catfish (January 31, September 2)
Colorado pikeminnow (February 20)
cutthroat trout (June 26)
Cyprinidae family (February 20)
darters (August 9)
egg casings (June 15)
lake trout (June 26)
mermaid's purses (June 15)
minnows (February 20)
paddlefish (April 8)
pike (January 31)
pupfish (October 8)
rainbow darter (August 9)
rays (June 15)
round goby (August 27)
salmon (May 18)
seahorses (October 9)
sharks (June 15)
shovel-nosed sturgeon (May 14)
skates (June 15)
snail darter (August 9)
steelhead (May 18)
striped bass (May 18)
sturgeon (May 18, September 26)
trout (January 31, August 18)
walleye (April 26)
whale shark (August 30)

FUNGI AND LICHEN

Amanita muscaria (November 24)
British soldier lichen (July 26)
chanterelles (October 25)
chicken of the woods (September 24)

chytrid fungus (May 8)
crustose lichens (July 26)
cryptobiotic crust (May 11)
fairy ring (August 24)
fly agaric (November 24)
foliose lichens (July 26)
fruticose lichens (July 26)
hen of the woods (September 24)
jack-o'-lantern (October 31)
lichens (April 18, July 26)
morel (March 24)
mycorrhizal fungi (September 18, October 25)
Oregon black truffle (February 23)
Oregon white truffle (February 23)
puffballs (June 23)
reindeer lichen (December 23)
stinkhorn (April 24)
truffles (February 23)

GEOLOGY

'a'ā (March 10)
aragonite (January 10)
caves (January 10)
Devonian Era (October 23)
fossils (October 13)
galena (August 10)
geothermal hot springs (July 12)
geysers (July 12)
glaciers (April 11)
Hawaiian islands (March 10)
hoodoos (November 13)
igneous rocks (August 11)
Jurassic period (August 29)
loess (November 8)
metamorphic rocks (June 8)
moraines, terminal and lateral (April 11)
obsidian (August 11)
pāhoehoe (March 10)
permafrost (July 11)
sandstone (December 2)
sedimentary rocks (December 2)
slate (June 8)
stalactites (January 10)
stalagmites (January 10)
thermokarst lakes (July 11)
Triassic period (September 26)

volcanoes (March 10)
vulcanization (July 31)

HABITAT

backyards (September 8)
backyard bird feeding (February 13)
bayous (February 7)
beaver dens (January 11)
beaver lodge (January 11)
circumpolar regions (March 19, July 25,
    December 23)
climax ecosystem (July 31)
coastal bays (May 9)
controlled-burns or prescribed fires (February 10)
cryptobiotic soil (May 11)
dreys (August 5)
ephemeral ponds or vernal pools (March 26)
estuaries (May 9)
fire, role of (February 10)
fynbos shrubland (April 10)
grasslands (August 14)
intertidal zone (July 9, November 27)
leks (March 30)
limnetic zone (November 1)
littoral zone (November 1)
mangrove roots (February 16)
mixed-grass prairies (August 14)
profundal zone (November 1)
rookeries (February 16, April 7)
sagebrush steppe (October 13)
shelter and space, importance of (February 9)
shortgrass prairies (August 14)
succession, primary and secondary (July 31)
tallgrass prairies (August 14, October 29)
tide pools (July 9)
wrack line (July 17)

HOLIDAYS AND CELEBRATIONS

Arbor Day (last Friday in April)
Bat Appreciation Month (October)
Bat Week (October)
Bike-to-Work Day (third Friday in May)
Bike-to-Work Week (May)
California Poppy Day (April 6)
Citizen Science Day (April 13)
Earth Day (April 22)

Earth Hour (end of March)
Earth Science Week (October)
Gopher Tortoise Day (April 10)
International Dark-Sky Week (April 1)
International Mountain Day (December 11)
International Whale Shark Day (August 30)
Latino Conservation Week (mid-July)
Manatee Awareness Month (November)
National Bike Month (May)
National Cherry Blossom Festival (April 6)
National Fossil Day (October 13)
National Moth Week (July 21)
Sea Otter Awareness Week (end of September)
sturgeon festivals (September 26)
Thanksgiving (November 25)
turkey vulture celebrations (September 23)
World Environment Day (June 5)
World Migratory Bird Day (May 10)
World Wetlands Day (February 2)

INVERTEBRATES

ants (April 10)
bald-faced hornet (January 28)
banana slug (January 14)
bark scorpion (July 20)
barnacles (April 18, July 9)
black-legged deer tick (February 28)
brine shrimp (June 20, August 8)
butterflies (June 14, July 2, August 28)
California spiny lobster (January 8)
centipedes (November 14)
cicadas, periodic (August 3)
crawdads (March 8)
crazy worms or jumping worms (June 7)
dung beetle (January 30)
fly (January 30)
earthworms (June 7)
emerald ash borer (December 14)
fairy shrimp (March 26)
freshwater mussels (May 14)
gall flies (October 2)
giant Pacific octopus (November 26)
ghost crab (March 18)
green darner (April 28)
hairstreak butterfly (December 26)
Halloween pennant dragonfly (October 31)

hawk or sphinx moth (May 20, July 2, September 12)
hickorynut mussel (May 14)
horseshoe crab (May 2, 3)
hummingbird moth (August 30)
Isabella tiger moth (November 30)
jumping spider (March 15)
ladybugs (September 28)
lightning bugs or fireflies (May 29, July 4, October 22)
limpets (July 9)
Maine lobster (January 8)
mayflies (June 28)
millipedes (November 14)
mollusks (August 2)
monarch butterfly (March 12, 14)
moon jelly (June 20)
Mormon cricket (July 24)
mosquitoes (June 2)
moths (July 2, December 31)
mourning cloak butterfly (February 11)
octopus (November 26)
paper wasps (January 28)
pimplebacks (May 14)
quagga mussels (May 14)
red-banded hairstreak butterfly (January 7)
rusty crayfish (March 8)
salmonflies (June 11)
sand dollars (December 27)
scorpions (July 20, August 21)
sea anemones (July 9)
sea cucumbers (December 27)
sea stars (July 9, November 27, December 27)
sea urchins (July 9, December 27)
signal crayfish (March 8)
snow fleas (December 28)
spiders, flying (July 29)
springtails (December 28)
stoneflies (June 11)
synchronous firefly (May 29)
ticks (February 28)
tarantula hawk (July 14)
tarantulas (July 14, September 14)
worm migration and vermicomposting (March 25)
yellowjackets (January 28)
yellow sandshell (May 14)
zebra mussel (May 14)
zebra swallowtail butterfly (September 17)

LEGISLATION

Alaska National Interest Lands Conservation Act
  (November 12, December 2)
Antiquities Act (June 8, June 10, September 24)
Civilian Conservation Corps (March 21)
Clean Air Act (April 22)
Clean Water Act (April 22, October 18)
Dingell-Johnson Act (September 2)
Endangered Species Act (March 11; June 9; August
  9, 16, 27; December 27)
Endangered Species Preservation Act (May 17)
hunting license, Pennsylvania (April 17)
National Environmental Policy Act (January 1)
Occupational Health and Safety Act (April 22)
Pittman-Robertson Act (September 2)
Reclamation Act (June 10)

MAMMALS

Arctic fox (July 25)
armadillo (July 22)
badger (August 10)
bats (March 1, June 24)
beavers (January 11, April 27, November 27)
beluga whale (August 1, November 6)
bighorn sheep (May 31, November 17, 23)
bison (February 26)
black bear (May 22)
black-footed ferret (May 17, September 13,
  November 27)
black-tailed prairie dog (March 7, September 13)
bowhead whale (August 2)
canids (January 20, May 31)
caribou (September 22; December 23, 24, 30)
cats, domestic (June 3)
Channel Island fox (October 18)
chimpanzees (February 5)
circumpolar wildlife (July 25)
coatis (July 5)
deer (May 31, June 10)
dogs, domestic (June 3)
elk (October 12)
felids (January 20, May 31)
fisher (February 19, October 30)
flying squirrels (February 23, March 1, August 5)
grasshopper mice (August 21)
gray fox (October 18)

gray squirrel (October 5)
groundhogs (February 2)
ground squirrels (January 1, August 5)
harbor seal (December 1)
hoary marmot (February 2)
gray whale (March 13)
javelina or collared peccary (April 20)
kangaroo rat (July 16)
lesser long-nosed bat (August 16)
long-nosed bat (June 24)
lynx (January 27)
manatees (December 29)
Mariana fruit bat (June 24)
marsupials (May 25)
moose (January 27, February 12, July 26)
mountain goat (May 31, November 23)
mountain lion, puma, cougar (February 19, June 10,
  October 5)
musk ox (July 25, December 9)
muskrat (April 27)
narwhals (November 6)
ocelot (September 16)
orca (August 1, November 6)
otters (April 5)
pallid bat (June 24)
pika (January 4)
pinnipeds (September 6)
polar bear (June 3, July 25)
porcupine (February 19, April 27)
prairie dogs (May 17, September 13, November 27)
pronghorn antelope (January 15, December 30)
ptarmigan (January 22)
rabbits (January 20)
raccoon (July 5)
red fox (October 18)
reindeer (September 22, December 23,
  December 24)
Richardson's ground squirrel (August 31)
ringtail cat (July 5)
river otters (April 5, May 27)
sea lions (September 6)
sea otters (April 5, end of September, November 27)
seals (September 6)
short-finned pilot whale (November 6)
shrews (April 15)
skunks (March 23)

snowshoe hares (January 22, 27)
star-nosed mole (November 8)
tree squirrels (August 5)
ungulates (October 12)
Virginia opossum (May 25)
walrus (May 5, July 25, September 6)
whales (August 1)
white-footed mice (September 30)
white weasel or ermine (January 9, 22)
wild boar (April 20)
wolverine (March 19)
wolves (January 12)
woodchucks (February 2)
yellow-bellied marmot (February 2)

NOTABLE PLACES
(NOT NECESSARILY PROTECTED)

Alaska (July 25)
Appalachian Mountains (June 8)
Appalachian Trail (September 20)
Arizona (April 29, July 5, October 7)
Arkansas (October 27)
Badlands, South Dakota (February 26)
Bahamas (June 6)
Barrow, Alaska (December 21)
Bayou Bartholomew (February 7)
Bay of Fundy (July 9, September 27)
Beaufort Sea (September 22)
Black River, North Carolina (November 2)
Blue Springs, Florida (December 29)
California (May 3, September 6)
Caloosahatchee River (June 10)
the Cascades (March 10)
Caspian Sea (September 26)
Central Park, New York City (March 6)
Chase Lake, North Dakota (January 24)
Chesapeake Bay (February 24)
Chihuahuan Desert (March 31)
Colorado (May 20, May 24)
Colorado Plateau (May 11)
Colorado River (February 20)
Columbia River (July 1)
Congress Avenue Bridge, Austin, Texas (June 24)
Crystal Geyser (July 12)
Crystal River (December 28)
Delaware (September 28, December 20)

Delaware Bay (May 3)
Eastern Egg Rock Island, Maine (July 13)
El Capitan (February 17)
the Everglades (December 7)
Fairbanks, Alaska (December 21)
Fire Fall. See Horsetail Fall
Flagstaff, Arizona (April 1)
Florida (January 3; March 14; September 3, 5, 12; October 6)
Florida Keys (February 21)
Grand Canyon (August 23)
Great Basin (May 11)
Great Lakes region (February 14, April 11, June 28, September 27)
Great Plains (February 26)
Great Salt Lake (August 8)
Green Mountains (June 8)
Greenland (July 25)
Hawai'i (October 8, November 7, December 19)
Homosassa River (December 29)
Horsetail Fall (February 17)
Indian River Lagoon (March 14)
Jamestown (July 30, August 6)
Kansas (February 26)
Lake Erie (August 27)
Lake Michigan (February 25)
Lake Minnetonka (April 14)
Lake Superior (October 15)
Land of the Standing-Up Rocks (April 18)
Long Valley Caldera (August 11)
Louisiana (March 4)
Lower Rio Grande Valley (August 28)
Marsh Lake, Minnesota (January 24)
Mill Grove, Pennsylvania (December 5)
Minnesota (June 27, September 25)
Mississippi (March 4)
Mississippi River (January 25, April 8, May 14, August 11, December 6)
Missouri River (September 7, November 8)
Mojave Desert (February 22, 27)
Mono Lake (August 8)
Mono-Inyo Craters (August 11)
Montana (February 11)
Mount Katahdin (July 8)
Mount Moran (May 27)
Mount St. Helens (August 19)

Murie Ranch, Wyoming (November 12)
Nevada (May 3, December 19)
New Mexico (May 14, August 17)
New Orleans (February 7)
New York City (March 6, September 11)
New Zealand (August 26)
North Cascades (November 23)
North Pole (July 25)
Ogasawara Islands, Japan (November 7)
Ohio (May 6)
Oklahoma (February 26, October 21)
Old Faithful (July 12)
Olympic Peninsula (November 23)
100th meridian (May 20, August 23)
Pennsylvania (April 17)
Pitchfork Ranch, Wyoming (May 17)
Platte River (March 16)
Port Clinton, Ohio (April 26)
Puerto Rico (January 17, July 4)
Rancho Nuevo Beach, Mexico (June 5)
Rio Grande (January 13, June 13)
Rio Grande Valley (January 3, September 16)
Saint John island (August 2)
Saint Petersburg, Florida (December 17)
Salt Lake City (July 24)
Salton Sea (January 24)
San Francisco Bay (February 24)
Snake River (May 27)
Soda Springs Geyser, Idaho (July 12)
Sonoran Desert (March 31, May 11)
South Carolina (January 3)
Steamboat Geyser, Yellowstone (July 12)
Texas (August 7, 28; September 16, November 9)
Tornado Alley (May 21)
Utah (July 24)
Veracruz, Mexico (October 11)
Washington, DC (April 6)
Whitnall Park, Wisconsin (June 7)
Wisconsin (August 10)
Witless Bay, Canada (July 13)
Wyoming (February 26, August 7)
Yellowstone Lake (January 24, June 26)
Yosemite Falls (February 17)

ORGANIZATIONS, ASSOCIATIONS, AND AGENCIES

Academy of Natural Sciences (September 7)
American Ornithological Society (May 6)
American Society of Ichthyologists and
    Herpetologists (July 18)
Appalachian Mountain Club (September 20)
Association of Zoos and Aquariums (May 7)
Bat Conservation International (June 24, October 2)
Bird Banding Lab (October 17)
Bird Studies Canada (February 15, December 25)
Bureau of Land Management, BLM. *See specific
    wildlife refuges under* Protected Places
California Academy of Sciences (October 28)
Citizen Science Association (April 13)
Cornell Lab of Ornithology (February 15, May 23,
    October 28)
Environment for the Americas (May 10)
FrogWatch USA (May 7)
Gopher Tortoise Council (April 10)
Hawkwatch International (October 11)
Hispanic Access Foundation (July 15)
International Astronomical Union (October 3)
International Dark-Sky Association (February 3,
    April 1)
Izaak Walton League (April 4)
Jane Goodall Institute (February 5)
Lady Bird Johnson Wildflower Center (April 17)
League of American Bicyclists (May 23)
The Mountaineers (July 8)
National Audubon Society (February 15,
    December 25)
National Aeronautics and Space Administration
    (July 3)
National Butterfly Center (August 28)
National Geographic Society (October 28,
    November 20)
National Hurricane Center (August 22)
National Park Service (February 17, August 25); *see
    also specific national parks, monuments, and recreation
    areas under* Protected Areas
National Phenology Network (January 25)
National Wildlife Federation (September 8)
New York City Audubon (September 11)
Project Seahorse (October 9)

Roger Tory Peterson Institute of Natural History (August 6)
Roots & Shoots (February 5)
Save the Manatee Club (November)
SciStarter (April 13)
Smithsonian's National Museum of Natural History (July 19)
US Environmental Protection Agency (March 9, June 18)
US Fish and Wildlife Service (October 14, December 29)
US Food and Drug Administration (November 29)
US Forest Service (October 4); *see also specific national forests under* Protected Places
US Geological Survey (September 19, October 17)
World Meteorological Organization (August 22)
Yale University School of Forestry (October 4)

PEOPLE

Abraham Lincoln (June 30)
Aldo Leopold (January 17)
Ansel Adams (April 22)
Archibald Menzies (June 13)
Archie Carr (May 21)
Bob Graham (November)
Carl Bergmann (December 11)
Carl Linnaeus (September 14)
Charles Darwin (May 1, July 29, August 29)
David Attenborough (May 8)
David Douglas (June 13)
Doris Mable Cochran (July 18)
Dwight Eisenhower (November 12)
E. O. Wilson (November 11)
Edward Abbey (January 29, April 20)
Eugene Schieffelin (March 6)
Franklin D. Roosevelt (March 21, September 2, October 14)
Galileo Galilei (February 15)
Gaylord Nelson (April 22)
George Bird Grinnell (December 5)
Gifford Pinchot (October 4)
Henry Foster (October 27)
Herbert Saffir (August 22)
Izaak Walton (April 4)
Jane Goodall (February 5)
Jared Potter Kirtland (June 6)

Jay Norwood "Ding" Darling (October 14, November 16)
Jimmy Buffett (November)
Jimmy Carter (November 12, December 2)
John James Audubon (October 16, December 5)
John Wesley Powell (February 20, August 22)
Joseph Asaph Allen (December 11)
Julius Caesar (February 29)
Lady Bird Johnson (April 17)
Loudon Wainwright III (March 23)
Lyndon B. Johnson (September 12)
Margaret "Mardy" Murie (November 12)
Margaret Morse Nice (May 6)
Marjory Stoneman Douglas (December 7)
Olaus Murie (November 12)
Pope Gregory XIII (February 29)
Rachel Carson (March 9)
Robert MacArthur (November 11)
Robert Simpson (August 22)
Roger Tory Peterson (July 30, August 6)
Dr. Seuss, a.k.a. Theodor Geisel (August 12)
Theodore Roosevelt (March 14; June 8, 10; July 1; October 4)
Thomas Nuttall (September 7)
Todd McGrain (November 5)
Truman Everts (September 4)
Ulysses S. Grant (March 1)
Walt Whitman (July 29)
Woodrow Wilson (August 25)

PROTECTED PLACES AND PUBLIC LANDS

Acadia National Park (February 26)
Allegany State Park (July 30)
Año Nuevo State Park (September 6)
Anza-Borrego Desert State Park (February 22)
Aransas National Wildlife Refuge (March 11)
Arches National Park (January 29)
Archie Carr National Wildlife Refuge (May 21, June 5)
Arctic National Wildlife Refuge (September 22, November 12)
Ashley National Forest (July 1)
Assateague Island National Seashore (January 7)
Badlands National Park (November 10)
Bayou Sauvage National Wildlife Refuge (February 7)

Big Bend National Park (June 12)

Big Cypress National Preserve (December 8)

Bighorn National Forest (February 22)

Black Canyon of the Gunnison National Park (February 3)

Black Hills National Forest (June 4)

Blue Ridge Parkway (June 17)

Bosque del Apache National Wildlife Refuge (January 13, December 22)

Bracken Cave Preserve (June 24)

Bryce Canyon National Park (November 13)

Canyonlands National Park (May 11, September 12)

Capitol Reef National Park (February 3)

Capulin Volcano National Monument (March 10)

Carlsbad Cavern National Park (May 14, June 24)

Chassahowitzka National Wildlife Refuge (March 11)

Chincoteague National Wildlife Refuge (January 7)

Chiricahua National Monument (April 18)

Colorado National Monument (May 24)

Conecuh National Forest (December 18)

Congaree National Park (May 29, October 22)

Corkscrew Swamp Sanctuary (September 12, November 2)

Craters of the Moon National Monument and Preserve (August 11)

Death Valley National Park (February 3, 22; October 31)

Denali National Park (February 26)

Deschutes National Forest (July 1)

Devils Tower National Monument (March 10, June 12, September 24)

Dinosaur Provincial Park (September 9)

Dinosaur Ridge, Morrison Fossil Area National Natural Landmark (September 9)

Dinosaur Valley State Park (September 9)

El Pinacate y Gran Desierto de Altar, Mexico (May 13)

El Yunque National Forest (January 17)

Everglades National Park (June 10; December 7, 8, 29)

Fishlake National Forest (September 29)

Fossil Butte National Monument (October 13)

Glacier National Park (April 11, July 8, July 30)

Glen Canyon National Recreation Area (May 11)

Gombe Stream National Park (February 5)

Grand Canyon National Park (February 26, June 12)

Grand Teton National Park (February 26; April 11; May 27; November 12, 23)

Great Basin National Park (May 11)

Great Smoky Mountains National Park (April 9, May 29, June 17)

Hartwick Pines State Park (June 6)

Hawai'i Volcanoes National Park (March 10)

Humboldt-Toiyabe National Forest (May 3)

Indiana Dunes National Park (February 25)

Indiana Dunes State Park (February 25)

Itasca State Park (December 6)

Jasper National Park (January 12)

Joshua Tree National Park (February 22, 27; October 31)

J. N. "Ding" Darling National Wildlife Refuge (October 14)

Kaibab National Forest (July 1)

Laguna Atascosa National Wildlife Refuge (September 16)

Lassen Volcanic National Park (June 12)

Loess Bluffs National Wildlife Refuge (November 8)

Loess Hills State Forest (November 8)

Mammoth Cave National Park (January 10)

Mendocino National Forest (July 1)

Midewin National Tallgrass Prairie (October 29)

Mojave National Preserve (February 22, May 11)

Mound Key Archaeological State Park (October 6)

Mount Rainier National Park (April 11, October 30)

Mount Rushmore National Memorial (June 12)

Muir Woods National Monument (June 12)

National Bison Range (February 26)

National Wildlife Refuge system (February 24)

Natural Bridges National Monument (June 12)

Neal Smith National Wildlife Refuge (October 29)

Necedah National Wildlife Refuge (March 11)

Nez Perce National Monument (July 1)

Olympic National Park (October 30)

Organ Pipe Cactus National Monument (May 13)

Padre Island National Seashore (June 5)

Pelican Island National Wildlife Refuge (March 14)

Pictured Rocks National Lakeshore (October 15)

Point Lobos State Natural Reserve (September 6)

Redwood National and State Parks (January 23)

Rio Grande Valley State Park (January 13)

Roan Mountain State Park (June 17)

Rocky Mountain National Park (September 29)

Shoshone National Forest (March 3)
Tallgrass Prairie National Preserve (October 29, November 12)
Tongass National Forest (May 22)
Tule Springs Fossil Beds National Monument (December 19)
Virgin Islands National Park (August 2)
Wheeler National Wildlife Refuge (March 11)
Wind Cave National Park (January 10)
Wood Buffalo National Park, Canada (March 11)
Yellowstone National Park (January 12; March 1; June 26, 30; July 12; September 4; November 23)
Yosemite National Park (February 17, June 30)
Zion National Park (November 19)

REPTILES

American alligator (September 5)
anole (November 7)
Burmese python (December 8)
collared lizard or mountain boomer (October 21)
desert tortoise (August 31)
freshwater turtles (January 31)
frogs (July 18)
geckos (July 7)
Gila monster (October 7)
gopher tortoises (March 3)
green anole (November 7)
green sea turtle (June 5)
hawksbill sea turtle (June 5)
horny toad or short-horned lizard (August 7)
indigo snake (December 18)
Kemp's ridley sea turtle (June 5)
Lake Erie watersnake (August 27)
leatherback sea turtle (June 5)
loggerhead sea turtle (June 5)
Microhylidae frogs (February 21)
northern watersnake (August 27)
sea turtles (May 21; June 5, 20; September 11)

snakes (July 18)
tortoises (March 3)
western banded gecko (July 7)

SCIENTIFIC RESEARCH AND CONCEPTS

bird banding (February 24, May 6)
climate change (July 11)
indicator species (January 4)
keystone species (November 27)
Lewis and Clark Expedition (January 15, July 30)
reintroduction (October 30)
taxonomic splits (July 10)
taxidermy (December 16)
turbidity (December 15)
vagrants (November 28)
water temperature (August 18)
wildlife management (January 17)

WEATHER PHENOMENA

clouds (October 24)
depth hoar (January 21)
dust devils (May 21)
fog (January 23)
haboobs or dust storms (July 27)
hoar frost (January 21)
hurricanes (August 22)
Hurricane Andrew (December 8)
ice-out (April 14)
lake effect snow (February 14)
ponds and ice (January 16)
Saffir-Simpson scale (August 22)
snow (January 9, 22; November 22)
snow caves (January 9)
tornadoes (May 21)
tropical storms (August 22)
water density (November 22)
watermelon snow (April 21)
waterspouts (May 21)

GET INVOLVED

These listings are organized by project, celebration, or organization name. Where applicable they include the sponsoring or founding organization name. Find a couple causes that spark your interest and pitch in.

All About Birds, www.allaboutbirds.org, Cornell Lab of Ornithology

American Ornithological Society, www.americanornithology.org

Arbor Day, Arbor Day Foundation, www.arborday.org

Association of Zoos and Aquariums, www.aza.org

Backyard Worlds: Planet 9, www.backyardworlds.org, NASA

Bat Conservation International, www.batcon.org

Bike-to-Work Day, Week, and Month, League of American Bicyclists, www.bikeleague.org

BioBlitz, www.nationalgeographic.org/projects/bioblitz, National Geographic

Bird Banding Lab, www.reportband.gov, Patuxent Wildlife Research Center

Budburst, https://budburst.org, Chicago Botanic Garden

California Poppy Day, www.wildlife.ca.gov/Conservation/Plants/California-Poppy

Christmas Bird Count, National Audubon Society, www.audubon.org, Bird Studies Canada, www.birdscanada.org

Citizen Science Day, Citizen Science Association, www.citizenscience.org, SciStarter, https://scistarter.org

Cool Green Science, https://blog.nature.org/science, Nature Conservancy

Crane Trust, https://cranetrust.org

Earth Day, Earth Day Network, www.earthday.org

Earth Hour, www.earthhour.org, World Wide Fund for Nature (a.k.a. World Wildlife Fund), www.worldwildlife.org

Earth Science Week, www.earthsciweek.org

eBird, www.ebird.org, Cornell Lab of Ornithology

FrogWatch USA, www.aza.org/frogwatch, Association of Zoos and Aquariums

Gopher Tortoise Day, Gopher Tortoise Council, http://gophertortoisecouncil.org

Great Backyard Bird, http://gbbc.birdcount.org

Hawkwatch International, https://hawkwatch.org

International Dark-Sky Week, International Dark-Sky Association, www.darksky.org

International Mountain Day, www.un.org/en/events/mountainday

iNaturalist, www.inaturalist.org, California Academy of Sciences and National Geographic

International Mountain Day, www.un.org/en/events/mountainday, United Nations

International Whale Shark Day, www.iucnredlist.org/species/19488/2365291

Izaak Walton League, www.iwla.org

Jane Goodall Institute, www.janegoodall.org

Journey North, www.journeynorth.org

Lady Bird Johnson Wildflower Center, www.wildflower.org

Latino Conservation Week, Hispanic Access Foundation, www.hispanicaccess.org

Manatee Awareness Month, Save the Manatee, www.savethemanatee.org

Mountain Watch, AT Seasons, and Mountain Watch View Guides programs, Appalachian Mountain Club, www.outdoors.org

National Butterfly Center, www.nationalbutterflycenter.org

National Cherry Blossom Festival, https://nationalcherryblossomfestival.org

National Fossil Day, www.nps.gov/subjects/fossilday/index.htm

National Moth Week, www.nationalmothweek.org

Nature's Notebook, National Phenology Network, www.usanpn.org

Project FeederWatch, www.feederwatch.org, Cornell Lab of Ornithology, www.birds.cornell.edu, Bird Studies Canada, www.birdscanada.org

Project NestWatch, https://nestwatch.org/

Roger Tory Peterson Institute, https://rtpi.org

Roots & Shoots, www.rootsandshoots.org

Sea Otter Awareness Week, www.seaotters.org/soaw

World Environment Day, www.worldenvironmentday.global

World Meteorological Organization, https://public.wmo.int/en

World Migratory Bird Day, www.migratorybirdday.org, Environment for the Americas, www.environmentamericas.org

World Wetlands Day, www.worldwetlandsday.org, Convention on Wetlands, www.ramsar.org

ACKNOWLEDGMENTS

Thanks to Heather Ray for the endless love and support, and additional kudos to you for being a keen editor who doesn't get too frustrated when I swap the terms *em dash* and *ellipsis*.

Special nod to my parents. I can't wait for another fishing trip with you, Pa. And, Ma, our adventures are always many of my favorites.

Numerous individuals helped brainstorm topics to include in the initial outline of the book. They also served as references for fact-checking. I'm grateful for your local, regional, and national perspectives: Scott Schaefer, Jarren Kuipers, Gwyn McKee, Patrick Hogan, Tim Feathers, Landy Figueroa, Clark Cotton, Uncle Larry, and Aunt Kathy.

Greg Russell and Karen Kerans were generous with their time and talents. Both contributed countless hours and innumerable edits to improve this text.

Rachel Scott offered up encouragement and support throughout the process.

I first met Kristina Polk when she was a talented young birder. She continues to inspire me and to keep me up to date on a number of conservation topics, especially marine mammals.

The spectrum of conservation entities is vast. State and federal agencies along with nonprofit organizations provide a solid foundation for both online learning and hands-on exploration. From species-specific efforts to broad-spectrum environmental topics, I turned to these as resources. Two sites I consult on a regular basis in my personal quest for knowledge are the Cornell Lab of Ornithology's All About Birds and the Nature Conservancy's Cool Green Science.

Thanks to Dr. Uwe Stender at Triada US Literary Agency for helping me navigate the ins, outs, and what-have-yous of publishing.

My membership in the Outdoor Writers Association of America has paid off yet again. Thanks to the group for helping me polish my skills as a communicator. This specific project was initiated at the organization's annual conference.

It has been nothing but a pleasure working with the top-notch professionals at Mountaineers Books, especially Kate Rogers and Laura Shauger. Copyeditor Erin Cusick cleaned up the manuscript while maintaining the voice, and I appreciate her efforts.

I'm proud of the words, but the stunning illustrations by Jeremy Collins are my favorite part of the book. It's an honor to share a book cover with an immensely gifted artist, especially with one who is also a champion for the planet.

Lastly, I'd like to acknowledge Yo-Yo Ma, Stuart Duncan, Edgar Meyer, and Chris Thile. The Goat Rodeo Sessions album was my soundtrack as I wrangled this book project together.

ABOUT THE AUTHOR AND ILLUSTRATOR

grandparents' place. One of his favorite adult memories is hiking the Thorofare of southeast Yellowstone and exploring the headwaters of Two Ocean Creek like Teddy Roosevelt did.

Learn more about Ken at www.kenkeffer.net.

KEN KEFFER is a naturalist and author with a background in wildlife biology. He has written eight books highlighting the importance of exploring the outdoors, including *The Kids' Outdoor Adventure Book*, which won a National Outdoor Book Award. As an educator, Ken uses nature as a classroom, teaching people the everyday lessons the outdoors offer. In 2015, he was named the Wisconsin Association for Environmental Education's Nonformal Educator of the Year. Ken has had a rich diversity of research and experiences, from monitoring small mammals in Grand Teton National Park to studying flying squirrels in Southeast Alaska to catching blue crabs off the coast of Maryland.

When he's not writing, Ken enjoys birding, floating on lazy rivers, and fly fishing. He and his wife Heather love to go camping with their dogs Willow the Wonder Mutt and Hazel the Wonder Nut. One of Ken's favorite childhood memories is building Fort Fishy along Rock Creek at his

JEREMY COLLINS roams the earth with sketchbooks in hand, pouring his soul into their pages. It is in the folds of those pages that his particular worldview was born—from authentic travel and adventures as an exploratory rock climber to award-winning filmmaker and author. Jeremy's passion is to explore wild destinations using art as medicine, conduit, and invitation to ask himself and others a universal question: how can art make the world better?

Follow Jeremy's adventures on Instagram at @jer.collins and online at jercollins.com.

SKIPSTONE is an imprint of independent nonprofit publisher MOUNTAINEERS BOOKS. It features thematically related titles that promote a deeper connection to our natural world through sustainable practice and backyard activism. Our readers live smart, play well, and typically engage with the community around them. Skipstone guides explore healthy lifestyles and how an outdoor life relates to the well-being of our planet, as well as of our own neighborhoods. Sustainable foods and gardens; healthful living; realistic and doable conservation at home; modern aspirations for community—Skipstone tries to address such topics in ways that emphasize active living, local and grassroots practices, and a small footprint.

Our hope is that Skipstone books will inspire you to effect change without losing your sense of humor, to celebrate the freedom and generosity of a life outdoors, and to move forward with gentle leaps or breathtaking bounds.

All of our publications, as part of our 501(c)(3) nonprofit program, are made possible through the generosity of donors and through sales of more than 700 titles on outdoor recreation, sustainable lifestyle, and conservation. To donate, purchase books, or learn more, visit us online:

www.skipstonebooks.org
www.mountaineersbooks.org

SKIPSTONE

LIVE LIFE

MAKE RIPPLES

YOU MAY ALSO ENJOY: